Biotech's

DICTIONARY *Of* Dairy Science

Compiled & Edited by
Dinesh Arora

First Indian Edition 2006

ISBN 81-7622-164-3

Published by : BIOTECH BOOKS
1123/74, Tri Nagar,
Delhi - 110 035
Phone: 27382765
e-mail: biotechbooks@yahoo.co.in

Showroom : 4762-63/23, Ansari Road, Darya Ganj,
New Delhi - 110 002
Phone: 23245578, 23244987

Printed at : Chawla Offset Printers
New Delhi - 110 052

PRINTED IN INDIA

Biotech's

DICTIONARY *Of* Dairy Science

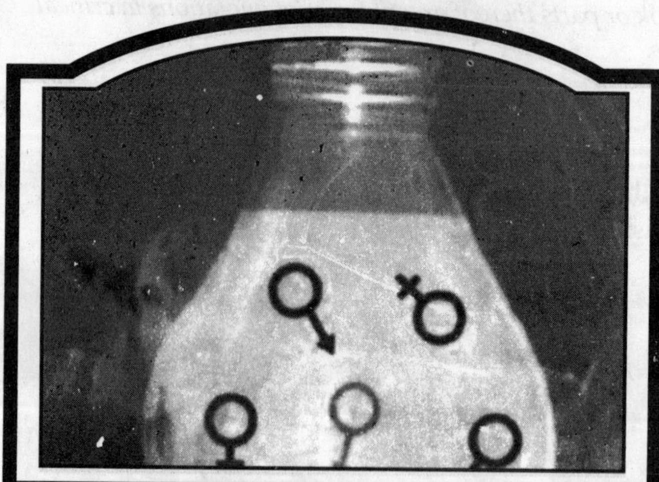

Compiled & Edited by

Dinesh Arora

BIOTECH

2006

BIOTECH BOOKS

DELHI - 110035

PREFACE

Animals have been the linchpin of growth and development of human civilisation and they will remain as such forever. It is known that domestication of animals was the most important step of history. Before learning agriculture man learned about animals and used milk, curd, hair and skin etc. of various animals according to his needs. This dependency was trans-formed to a new relationship after the advent of agriculture. Now animals were used for activities like ploughing, transportation, Sports and entertainment, sacrifices etc. This change, unknowingly though, gave a new turn to man-animal relationship. Now man started knowing about various species of animals. Further studies pertaining to anatomical and morpholo-gical structures of animals were made. This gave birth to field of animal studies or veterinary and dairy science. As the knowledge about usages, diseases, cure and medicines increased, their proper recording became a compulsory activity, which had been done from time immemorial.

The unceasing need of further discovery, innovation, reviving of old and relevant methods for proper mainte-nance of livestock, or to increase dairy production etc, becomes inevitable today because this is an important part of primary economic activity. And it will be equal to showing light to sun, as to say that about one third of world population is still dependant on primary economic activities.

Therefore, present dictionary is brought forward as a compilation to satiate the needs of students and teachers alike. The dictionary uses simple language to explain the intricate and confusing jargon of the subject, which is supported by real life examples and appropriate pictures, giving the readers an enriching learning experience. Utmost care has been taken to record as much information with its full explanations as possible.

■ **abdomen**
a region of the body. In vertebrates this is the part of the body that contains all the internal organs except for the heart and lungs. In most arthropods this is the hind region of the body.

■ **abnormalities**
unwanted or undesirable features.

■ **abomasum**
the fourth stomach compartment (true stomach) of a ruminant.

■ **abortion**
premature expulsion of the fetus or unborn animal.

■ **absorbent packing**
material within a package which absorbs liquids from product, pad in meat trays is made from paper and has a plastic liner.

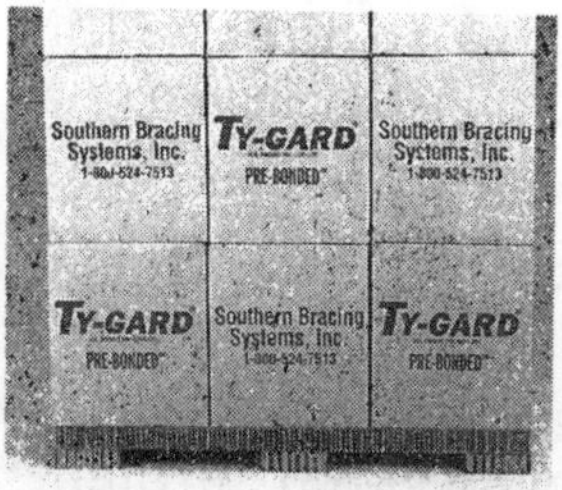

■ **absorptive hypercalciuria**
abnormally high urinary calcium caused by one of several mechanisms that increase calcium absorption from the digestive tract. Type I is the most severe and doesn't respond well to dietary calcium moderation which works well in Type II. The most common cause of high urinary calcium.

■ **abstraction**
removal of water from a watercourse or waterbody (e.g. for use in industry).

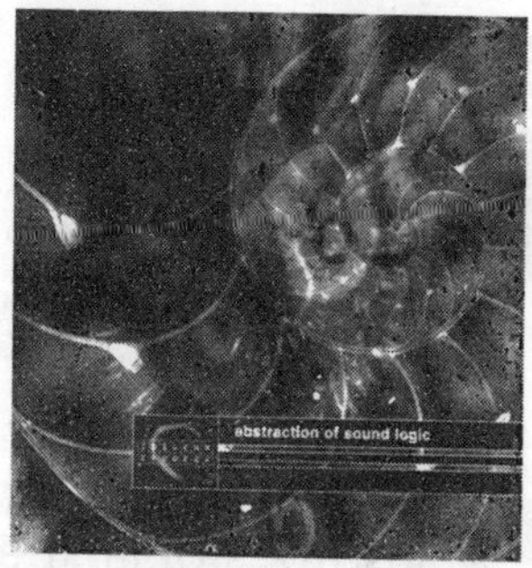

■ **AC**
the abbreviation for the chemotherapy drug combination of an anthracycline (usually doxorubicin) and cyclophospharnide.

■ **acclimatisation**
adaptation of an organism to a new environment.

■ **accretion**
gradual increase in size through addition of material.

■ **acetonemia (ketosis)**
a condition characterised by an elevated concentration of ketone bodies in body tissues and fluids. It is more common among high-producing cows in a negative energy balance.

■ **acid**

1. a sour substance with corrosive power. Acid substances have pH values of less than 7.
2. any chemical that forms chemical reactions by releasing hydrogen. In practical terms, this often destroys or digests the receiving compound depending on the strength of the acid. The most common biological acids are stomach acid and uric acid. Vinegar and lemon juice are other examples. Acids general produce a sour taste.

■ **acid coagulation**
acidification causes the casein micelles to destabilise or aggregate by decreasing their electric charge to that of the isoelectric point. At the same time, the acidity of the medium increases the solubility of minerals so that organic calcium and phosphorus contained in the micelle gradually become soluble in the aqueous phase. Casein micelles disintegrate and casein precipitates. Aggregation occurs as a result of entropically driven hydrophobic interactions.

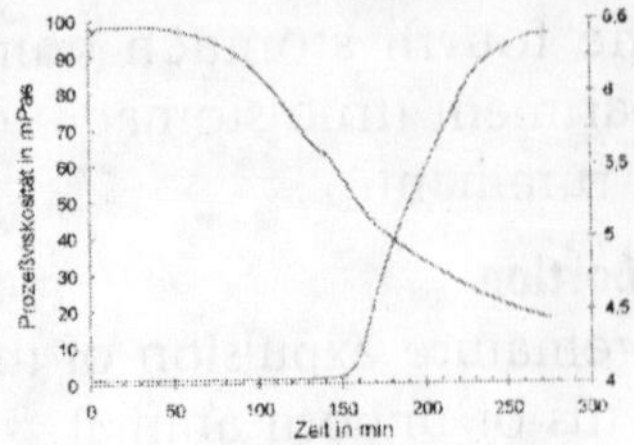

■ **acid detergent fibre**
fibre measurement extracted with acidic detergent in a technique employed to help appraise the quality of forages. Includes cellulose, lignin, ADIN and acid-insoluble ash.

■ **acid rinse**
part of the equipment cleaning process for stainless steel and rubber parts, removes fat, protein and minerals and also reduces bacteria. (See Equipment Sanitisation)

- **acid-base equilibria**
  both titratable acidity and pH are used to measure milk acidity. The pH of milk at 25° C normally varies within a relatively narrow range of 6.5 to 6.7. The normal range for titratable acidity of herd milks is 13 to 20 mmol/L. Because of the large inherent variation, the measure of titratable acidity has little practical value except to measure changes in acidity (e.g. during lactic fermentation) and even for this purpose, pH is a better measurement.
  There are many components in milk which provide a buffering action. The major buffering groups of milk are caseins and phosphate.

- **acidic**
  pertaining to or containing acid.

- **acidification**
  changes in chemistry in an area that lead to a more acidic environment (e.g. acid rain).

- **acidity test**
  milk quality test that measures lactic acid.

- **acidophilus**
  important bacteria found naturally in the human small intestine. Lactic acid bacteria L.acidophilus are used to ferment foods and to make yogurt.

- **acidosis**
  condition where the body has a higher acid load than normal. Urinary citrate excretion may probably be reduced in this situation. Treatment is usually with supplemental citrate or bicarbonate.

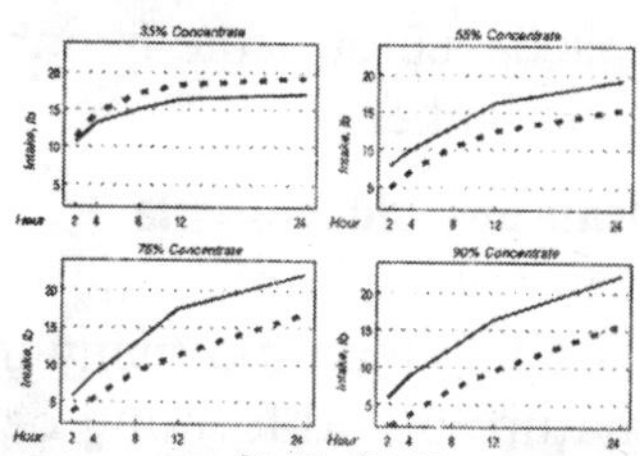

- **active immunity**
  a type of acquired immunity whereby resistance to a disease is built up by either having the disease or receiving a vaccine to it.

■ **aculeate**

prickly or stinging. For example, 'aculeate hymenoptera' are insects of the division hymenoptera that have stings (bees, ants and wasps).

■ **acute**

1. having a short and relatively severe course, not chronic.
2. used to describe disease where symptoms are readily evident. Treatment is generally required.

■ **acute lymphoblastic leukemia**

a quickly progressing disease in which too many immature white blood cells called lymphoblasts are found in the blood and bone marrow. Also called ALL or acute lymphocytic leukemia.

■ **acute myeloid leukemia**

a quickly progressing disease in which too many immature blood-forming cells are found in the blood and bone marrow.

Promyelocytic leukemia is a type of acute myeloid leukaemia. Also called AML or acute myelogenous leukemia.

■ **acyclovir**

a drug used to treat viral infections. Also known as Zovirax.

■ **ad libitum (ad lib.)**

at pleasure. Commonly used to express feed available on free-choice basis.

■ **adaptation**

a change in a characteristic of an organism that makes it better suited to survive in a particular environment.

■ **adenovirus**

a group of viruses that cause respiratory tract and eye infections. Adenoviruses used in gene therapy are altered to carry a specific tumour-fighting gene.

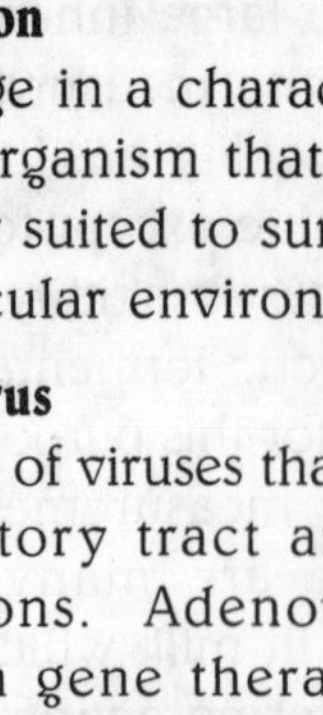

■ **adin acid**

detergent insoluble nitrogen. Undigestible dietary protein or nitrogen.

- **adjuvant**
  insoluble material that increases the formation and persistence of antibodies when injected with an antigen.

- **adjuvant therapy**
  adjuvant therapy is treatment given in addition to the primary treatment. Such therapy may be given to increase the effectiveness of the primary treatment or to prevent the spread or recurrence of cancer, or may consist of the use of a substance to enhance the body's immune response. Adjuvant therapy may be chemotherapy, radiation therapy or hormone therapy.

- **adulterated food**
  generally, impure, unsafe, or unwholesome; however, the Federal Food, Drug and Cosmetic Act, the Federal Meat Inspection Act, the Poultry Products Inspection Act, and the Egg Products Inspection Act contain separate language defining in very specific (and lengthy) terms how the term 'adulterated' will be applied to the foods each of these laws regulates. Products found to be adulterated under these laws cannot enter into commerce for human food use.

- **adulteration**
  addition of other substances to milk which reduces the quality of the milk (e.g. water)

- **aerobic**
  needing oxygen for growth.

- **aerobic requiring oxygen**
  many microorganisms require oxygen for the oxidation of food materials.

- **afforestation**
  planting of large areas with trees.

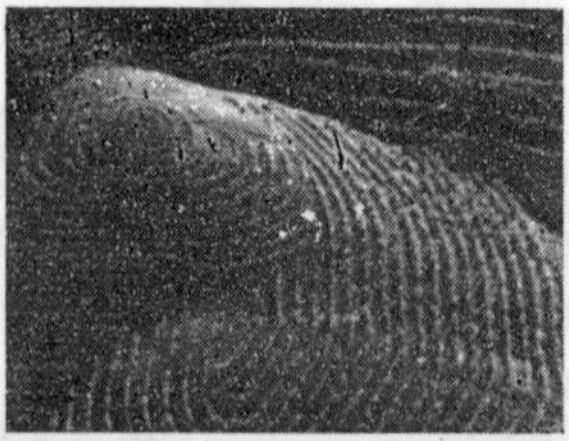

■ **afterbirth**
the placenta and allied membranes with which the fetus is connected with the mother. It is expelled from the uterus following parturition.

■ **ag337**
an anticancer drug used to shrink tumours, may also enhance the effectiveness of radiation therapy.

■ **agalactia**
failure to secrete milk following parturition.

■ **age and month-of-calving factors**
factors used to eliminate the environmental effects of different ages and months of the year at calving. These factors standardise lactation records for genetic evaluations.

■ **age at last calving**
age determined by subtracting the cow's birth date from her most recent calving date.

■ **age gelation**
it is an aggregation phenomenon that affects shelf-stable, sterilised dairy products, such as concentrated milk and UHT milk products. After weeks to months storage of these products, there is a sudden sharp increase in viscosity accompanied by visible gelation and irreversible aggregation of the micelles into long chains forming a three-dimensional network. The actual cause and mechanism is not yet clear, however, some theories exist:
Proteolytic breakdown of the casein: bacterial or native plasmin enzymes that are resistant to heat treatment may lead to the formation of a gel
Chemical reactions: polymerisation of casein and whey proteins due to Maillard type or other chemical reactions. Formation of kappa-casein-ß -lactoglobulin complexes.
An excellent source of information on casein micelle stability can be found in Walstra.

- **aggressive lymphoma**
  lymphoma that grows quickly.

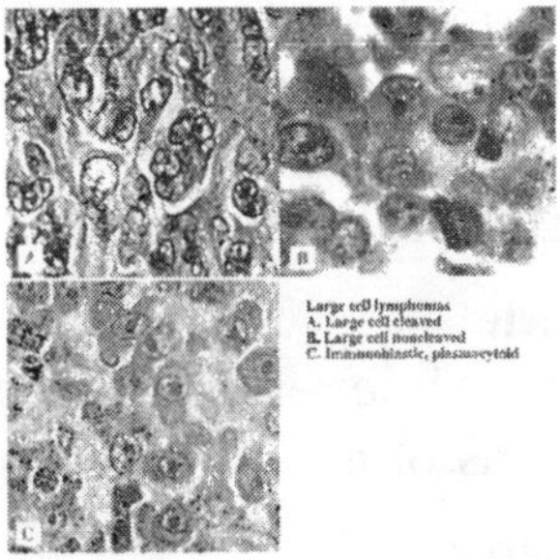

- **agricultural intensification**
  the movement towards increasing farm production by use of agrochemicals and mechanisation.

- **agri-environment**
  relating to the incentive payments that are available to farmers to voluntarily pursue less intensive forms of agriculture that are of benefit to the environment.

- **agrobacterium tumefaciens**
  a common soil bacterium used as a vector to create transgenic plants.

- **agrochemical**
  synthetic chemicals used in modern agriculture such as chemical fertilisers, herbicides, pesticides etc.

- **AI daughter**
  female offspring of an artificially inseminated dam.

- **albumin**
  any protein that is soluble in water.

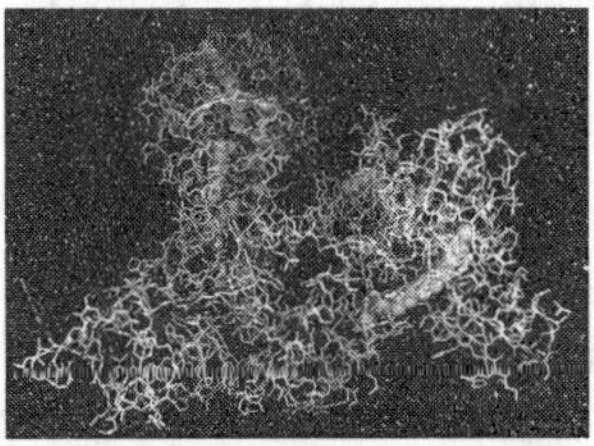

- **alcohol test**
  milk quality test.

- **alien species**
  a species that does not normally occur in an area but which has been deliberately or accidentally introduced by human activity (e.g. Rhododendron planted by the Victorians in many Lancashire woodlands as cover for game).

- **alkali**
  highly soluble in water producing caustic or corrosive solutions that neutralise ac-

ids. Alkaline substances have pH values above 7.

- **alkaline or alkili**

another name for base or antacid, the opposite of acidic.

- **alkaline phosphatase**

phosphatase enzymes are able to split specific phosporic acid esters into phosphoric acid and the related alcohols. Unlike most milk enzymes, it has a pH and temperature optima differing from physiological values; pH of 9.8. Minimum pasteurisation temperatures destroy the enzyme, therefore, a phosphatase test can be done to ensure proper pasteurisation.

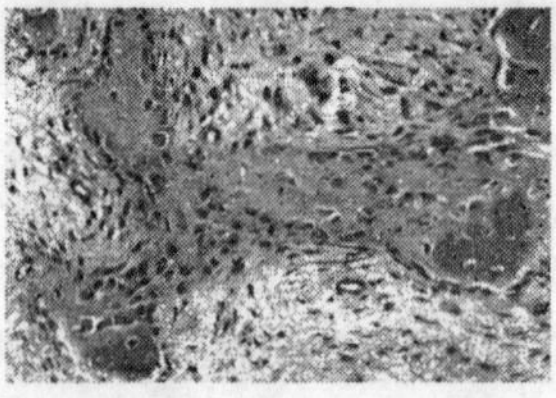

- **alkaloids**

drugs used in chemotherapy. Vinca alkaloids can kill cancer cells.

- **alkylating agents**

a family of anticancer drugs (including nitrosoureas) that combine with a cancer cell's DNA to prevent normal cell division. Alkylating agents include busulfan, carmustine, carzelesin, cyclophosphamide (also called Cytoxan), ifosfamide, lomustine, melphalan, porfiromycin and semustine.

- **allele**

any of several alternative forms of a gene.

- **allergy**

hypersensitive reaction to intrinsically harmless antigens, most of which are environmental. Allergies are divided into those that produce immediate or antibody-mediated reactions and those that produce delayed or cell-mediated reactions. Immediate reactions release certain substances into the circulation, such as histamine, bradykinin, acetylcholine, immunoglobulin IgG, and leukotaxine. Delayed reactions are caused by antigens, but do not seem to depend on antibodies. Some common symptoms of allergy are bronchial congestion, conjunctivitis, edema, fever, urticaria, and vomitting. Severe reactions include anaphylaxis and angioneurotic edema of the glottis.

- **alley**
  a walking area for cattle within a barn such as a loafing alley, feeding alley or cross alley (walkway) from a barn to the milking parlour.

- **alley**
  a walking area for cattle within a barn, as a loafing alley or feeding alley.

- **alley scraper**
  a 'V' shaped mechanical blade that is dragged over an alley by chain or cable to pull manure to collection channel at the end of the alley (or possibly the centre of the barn). The blade then collapses and is drawn back to the opposite end of the alley.

- **allogeneic**
  adjective referring to the transfer of material such as bone marrow or skin from one person (the 'donor') to another person ('the recipient').

- **allogenic**
  of the same species, but with a different genotype. Also allogeneic.

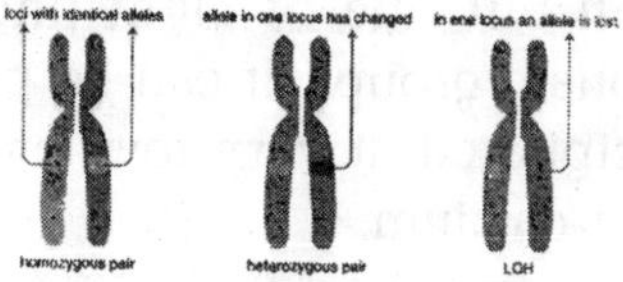

Schematic rendition of a pair of homologous chromosomes under conditions of homozygosity, heterozygosity and LOH. Homologous chromosomes are those pair during mitosis, have the same morphology, and contain genes governing the same characteristics. Note that one of the chromosomes, say the blue one, came from the father, and the other, the red, from the mother.

- **allopurinol**
  a drug that lowers high uric acid (a byproduct of metabolism) levels in the blood caused by some cancer treatments or by gout.

- **allovectin-7**
  a compound used for gene therapy.

- **alopecia**
  the loss of hair from the body and/or scalp. An effect of some cancer treatments.

- **alopecia baldness**
  loss or deficiency of hair, natural or abnormal.

- **alpha(s1)-casein**
  two hydrophobic regions, containing all the proline residues, separated by a polar region, which contains

all but one of eight phosphate groups. It can be precipitated at very low levels of calcium.

- **alpha(s2)-casein**

concentrated negative charges near N-terminus and positive charges near C-terminus. It can also be precipitated at very low levels of calcium.

- **alpha-lactalbumins**

these proteins contain eight cysteine groups, all involved in internal disulfide bonds, and four tryptophan residues. alpha-Lactalbumin has a highly ordered secondary structure, and a compact, spherical tertiary structure. Thermal denaturation and pH <4.0 results in the release of bound calcium.

- **altretamine**

an chemotherapy drug that interferes with the growth of tumour cells and kills them.

- **alzheimer's disease**

a disease characterised by, among other things, progressive loss of memory. The development of alzheimer's disease is thought to be associated, in part, with possessing certain alleles of the gene that encodes apolipoprotein E.

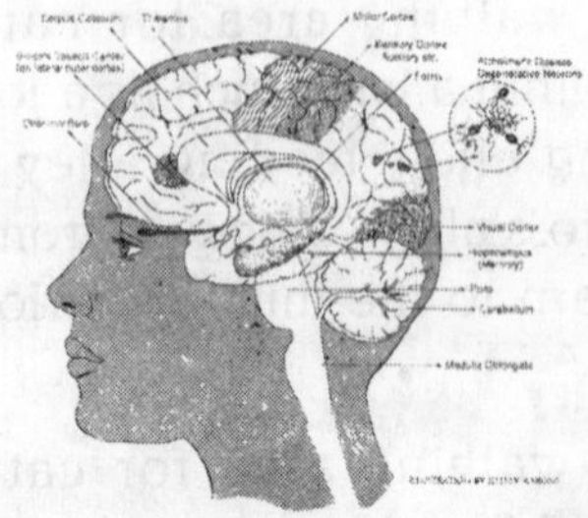

- **amenity planting**

planting trees and shrubs for non- commercial purposes, for example, planting trees not primarily as a crop but for landscaping reasons.

- **amifostine**

a drug used to control some side effects of chemotherapy and radiation therapy.

- **amino acid**

one of the building blocks of protein. Formed by the combination of a carbohydrate (sugar) and a nitrogen compound.

- **amino acids**

1. building blocks of proteins. There are 20 common amino acids: alanine, arginine, asparg-ine, aspartic acid, cysteine, glutamic acid, glutamine, glycine,

histidine, isoleucine, leucine, lysine, methionine, phenylalanine, proline, serine, threonine, tryptophan, tyrosine and valine. Two more amino acids have been discovered in microbes: selenocysteine and pyrrolysine.

2. organic compounds linked together in a linear chain to form proteins. There are 20 amino acids commonly found in proteins.

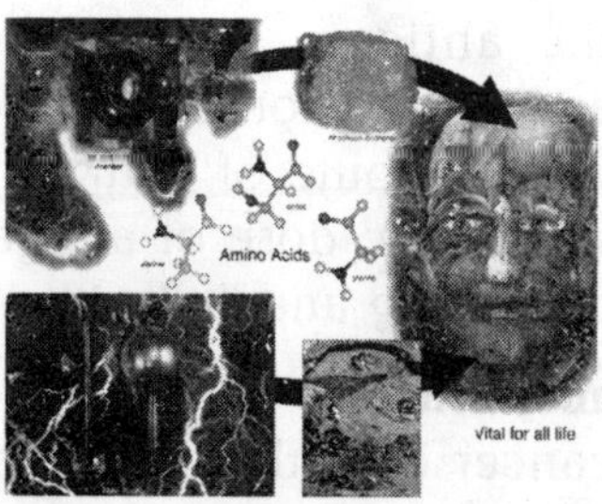

- **aminocamptothecin**
  belongs to the group of drugs known as topoisomerase inhibitors. Also known as 9-AC.

- **aminoglutethimide**
  a drug used to decrease hormone production and suppress tumour growth. Also known as Cytadren.

- **ammonia**
  a colourless pungent gas. NH3, composed of nitrogen and hydrogen; its compounds are used as fertilisers.

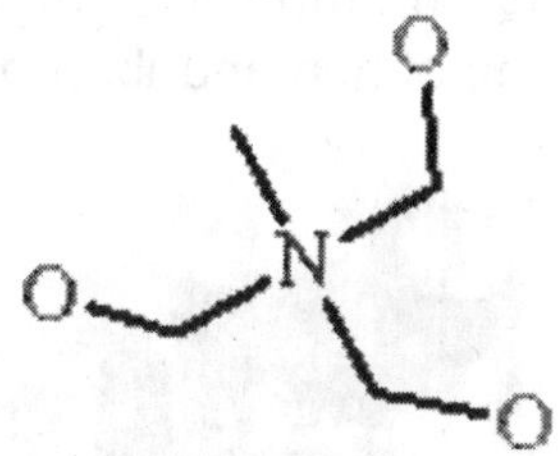

- **ammonium**
  an ion (NH4+) derived from ammonia (NH3).

- **amplification**
  the process of increasing the number of copies of a particular gene or chromosomal sequence.

- **anaemia**
  a condition in which the number of red blood cells is below normal.

- **anaerobic**
  growing in the absence of oxygen.

- **anaerobic treatment lagoon**
  a structure to treat animal waste by predominantly anaerobic biological action using anaerobic or facultative organisms, in the absence of air, for the purpose

of reducing organic matter in wastes.

- **analgesics**
drugs that reduce pain. These drugs include aspirin, acetaminophen and ibuprofen.

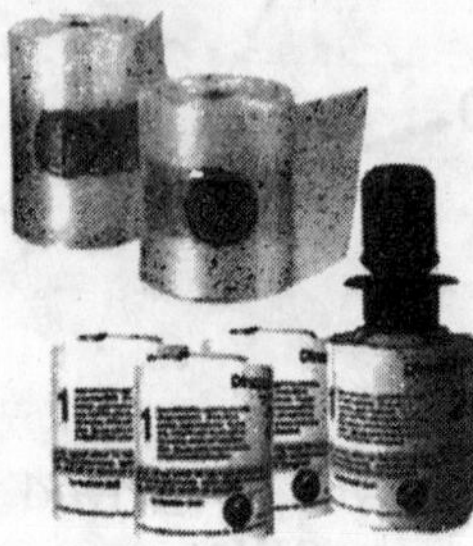

- **analogue**
a chemical compound with a similar structure to another but different from it in some respect. May have a similar or opposite effect in the body.

- **analyse**
to look at, to identify, to assess.

- **anaphylactic shock (anaphylaxis)**
a state of immediate hypersensitivity following sensitisation to a foreign protein or drug.

- **anaphylaxis**
an exaggerated, life-threatening hypersensitivity reaction to a previously encountered antigen. The response, which is mediated by antibodies of the IgE class of immunoglobulins, causes the release of the chemical mediators from the mast cells. The reaction may be a localised wheal and flare of generalised itching, hyperemia, angioneurotic edema, and in severe cases vascular collapse, bronchospasm, and shock. The severity of symptoms depends on the original sensitising dose of the antigen, the amount and distribution of antibodies, the route of entry and size of the dose of antigen producing anaphylaxis.

- **anaplastic**
cancer cells that grow and divide rapidly.

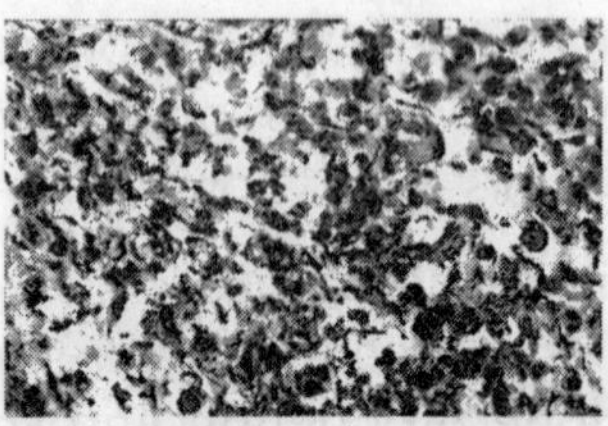

- **anaplylaxis**
a severe allergic reaction in which the breathing tube or irway can swell and shut off the air supply ultimately

cause death if not treated rapidly. Epinephrine (adrenalin) is the treatment of choice.

- **anastrozole**
a drug used to prevent the formation of estrogen from other hormones.

- **ancient woodland**
woodland that has existed since at least 1600 AD.

- **androgen suppression**
treatment to suppress or block the production of male hormones in order to reduce the spread of prostate cancer. Androgen suppression is achieved by removing the testicles, by taking female sex hormones, or by taking drugs. Also called androgen ablation.

- **anesthesia**
condition produced in order to permit a painless surgical operation. Loss of sensation occurs.

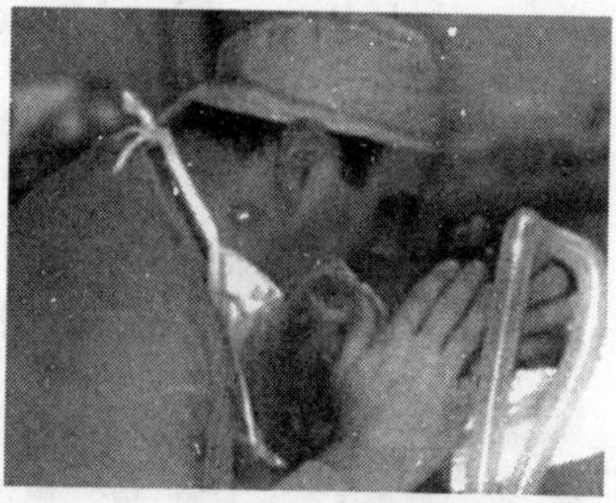

- **anesthesiologist**
a physician who specialises in the administration of nesthesia.

- **anetholtrithione**
a drug that may prevent the development or progression of cancer.

- **anhydrous**
containing no water.

- **animal food**
any article intended for use as food for dogs, cats, or other animals derived wholly, or in part, from the carcass or parts or products of the carcass of any livestock, except that the term animal food as used herein does not include:
1. Processed dry animal food or
2. Livestock or poultry feeds manufactured from processed livestock byproducts (such as meatmeal tankage, meat and bonemeal, bloodmeal, and feed grade animal fat).

- **animal food manufacturer**
any person engaged in the business of manufacturing or processing animal food.

■ **animal health care services**
services provided by the group to prevent and treat sick animals.

■ **animal unit**
a unit of measurement of livestock, the equivalent of one mature cow weighing 1,000 pounds. The measure is used in making comparisons of feed consumption and grazing.

■ **annual**
a plant that completes its life cycle in a single growing season.

■ **anorexia**
lack or loss of appetite.

■ **anthelmintic**
a remedy for destruction or elimination of parasitic worms.

■ **anthracenediones**
a group of drugs used in chemotherapy. This group includes mitoxantrone.

■ **anthraquinones**
a group of chemotherapy drugs.

■ **anthrax**
a disease of mammals and humans caused by a spore-forming bacterium called Bacillus anthracis. Anthrax has an almost worldwide distribution and is a zoonotic disease, meaning it may spread from animals to humans. All mammals appear to be susceptible to anthrax to some degree, but ruminants such as cattle, sheep, and goats are the most susceptible and commonly affected, followed by horses and then swine.

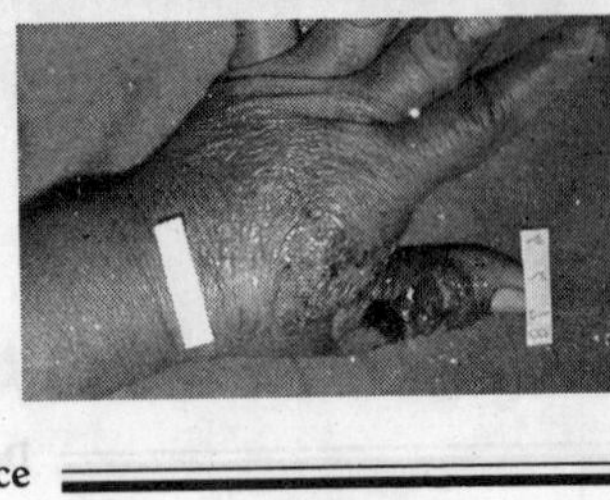

- **antiandrogen**
  drugs used to block the production of male hormones. Antiandrogens include bicalutamide, cyproterone acetate and flutamide (Eulexin).

- **antiandrogen therapy**
  treatment with drugs to block the production of male hormones.

- **antiangiogenesis**
  prevention of the growth of new blood vessels to a tumour.

- **antibiotic**
  a substance that destroys or inhibits the growth or action of microorganisms.

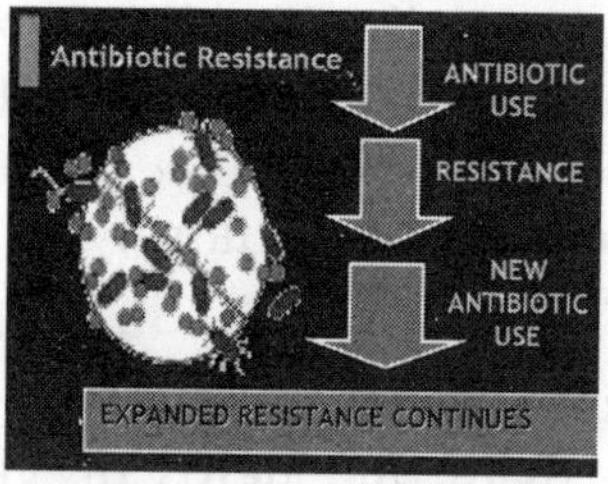

- **antibiotic**
  1. chemical substance formed as a metabolic byproduct in bacteria or fungi and used to treat bacterial infections. Antibiotics can be produced naturally, using microorganisms or synthetically.
  2. a metabolic product of one microorganism or a chemical that in low concentrations is detrimental to activities of specific other microorganisms. Examples include penicillin, tetracycline, and streptomycin. Not effective against viruses. A drug that kills microorganisms that cause mastitis or other infectious disease.
  3. a metabolic product of one microorganism or a chemical that in low concentrations is detrimental to activities of other microorganisms. Penicillin, tetracycline, and streptomycin are antibiotics. Not effective against viruses.

- **antibiotic residues**
  the presence of traces of antibiotics or their derivatives in milk or meat.

- **antibiotic test kit**
  test kit for use on the farm to detect residues of antibiotics in milk before the milk is picked up for delivery to the plant.

- **antibiotics**
  drugs that fight infection caused by bacteria. Antibiotic drugs include amikacin,

amoxicillin (amoxicillin-clavulanic acid), ceftriaxone, ciprofloxacin, clarithromycin, imipenem, metronidazole, novobiocin, penicillin, tetracycline and trimethoprim-sulfameth-oxazole.

- **antibodies**
  1. protect young mammal from infection, secreted into milk, especially colostrums milk.
  2. disease-fighting substances produced by certain white blood cells and found in the blood. Antibodies interact only with a specific target (antigen). Also known as 'immunoglobulins'.
  3. proteins produced by an organism's immune system that recognise foreign proteins (or their surface sugars). Specific antibodies can be produced in large volumes and employed to detect, diagnose, or treat one or more diseases.

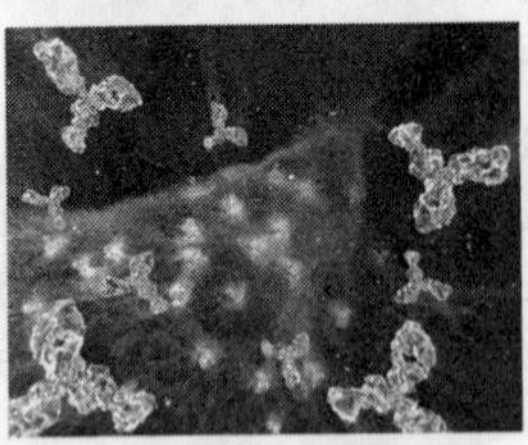

- **antibody**
  1. protein produced by humans and higher animals in response to the presence of a specific antigen.
  2. a protein (modified type of blood serum) developed or synthesised by lymphoid tissue of the body in response to an antigen. Each antigen elicits production of a specific antibody.

- **antibody therapy**

  treatment with an antibody, a substance that can directly kill specific tumour cells or stimulate the immune system to kill tumour cells.

- **anti-cea antibody**

  an antibody developed against carcinoembryonic antigen (CEA), a protein present on certain types of cancer cells.

- **anticoagulant**

  a drug that helps prevent blood clots from forming. Also called a blood thinner.

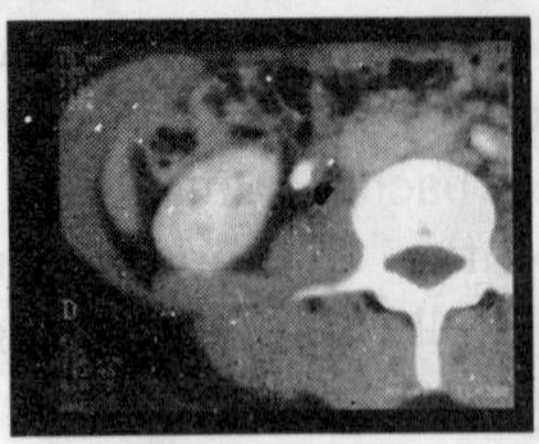

■ **anticodon**

triplet of nucleotide bases (codon) in transfer RNA that pairs with (is complementary to) a triplet in messenger RNA. For example, if the codon is UCG, the anticodon is AGC.

■ **anticonvulsant**

a drug that prevents or relieves convulsions or seizures.

■ **antiemetic**

a drug that prevents or reduces nausea and vomiting that may be associated with anticancer therapies. Antiemetic drugs include granisetron, metoclopramide and ondansetron.

■ **antifungal**

a drug that treats infections caused by fungi. Antifungal drugs include amphotericin B, fluconazole, flucytosine, metronidazole, mycostatin, nystatin and voriconazole.

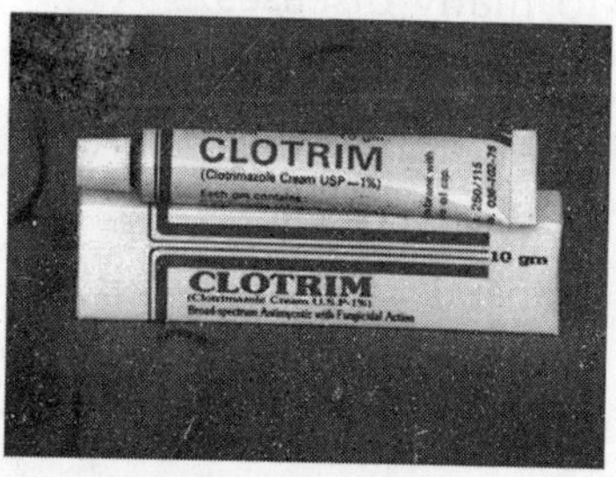

■ **antigen**

1. a substance that is recognised by the immune system as foreign to the body.
2. a high-molecular-weight substance (usually protein) which, when foreign to the bloodstream of an animal, stimulates formation of a specific antibody.

■ **antihemophilic factors**

a family of whole-blood proteins that initiate blood clotting. Some of these proteins, such as factor VIII, can be used to treat haemophilia. See **factor viii**, **kidney plasminogen activator**.

■ **antimetabolites**

a group of chemotherapy drugs that stop cancer from growing. Antimetabolites include cladribine (2-chlorodeoxyadenosine), capecitabine, cytarabine (also called ARA-C and cytosine arabinoside), edatrexate, fluorouracil (5-FU), methotrexate, pentostatin, piritrexim, and trimetrexate.

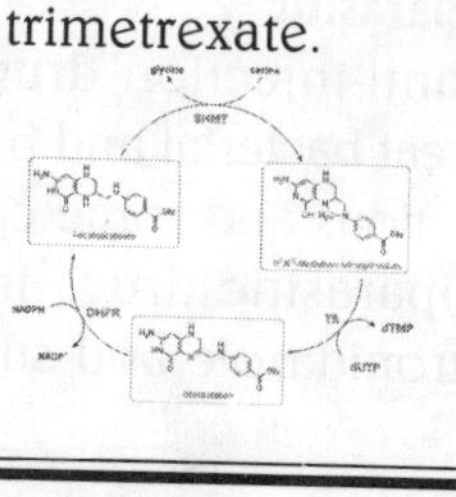

- **antimicrobial resistance**

  bacteria and other disease-causing organisms have a remarkable ability to mutate and acquire resistance genes from other organisms and thereby develop resistance to antimicrobial drugs. When an antimicrobial drug is used, the selective pressure exerted by the drug favours the growth of organisms that are resistant to the drug's action.

- **antineoplastic antibiotics**

  a group of chemotherapy drugs including cordycepin and the anthracyclines doxorubicin (Adriamycin, AD 32), daunorubicin, epirubicin and mitomycin.

- **antioxidant**

  substance added to food to prevent the oxygen present in the air from causing undesirable changes in flavour colour. BHA, BHT and tocopherols are examples of antioxidants.

- **antiparasitic**

  an anti-infection drug used to treat bacterial and parasitic infections and some cancers. Antiparasitic drugs include metronidazole and surainin.

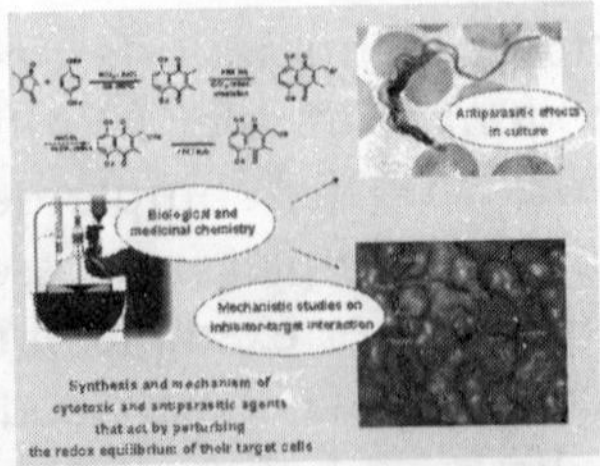

- **antisense**

  a piece of DNA producing a mirror image 'antisense' messenger RNA that is opposite in sequence to one directing protein synthesis. Antisense technology is used to selectively turn off production of certain proteins.

- **antisense c-fos**

  synthetic genetic material that may slow or stop the growth of cancer cells, being tested for use as gene therapy.

- **antiserum**

  blood serum containing specific antibodies against an antigen. Antisera are used to confer passive immunity to many diseases.

- **antithymocyte globulin**

  a substance that increases immune responses.

- **antiviral**

  a drug or substance active against viruses, Viruses are

submicroscopic organisms causing infectious disease. Antiviral drugs include cidofovir, didanosine, stavudine, lamivudine and indinavir.

- **aphid**
  a type of soft- bodied insect that feeds on plant sap. Ants that feed on the honeydew the aphids produce tend many species. Greenfly and blackfly are types of aphid.

- **apolipoprotein E (apo E)**
  certain alleles of the gene that encodes the protein apolipoprotein E have been associated with the development of heart disease and alzheimer's disease.

- **aquaculture**
  the production of aquatic plants or animals in a controlled environment, such as ponds, raceways, tanks or cages, for all or part of their life cycle.

- **arable**
  farming system in which crop plants (e.g. cereals) are raised rather than livestock.

- **arable stewardship**
  an agri-environment grant scheme that has been piloted in East Anglia and Shropshire, providing incentives to arable farmers.

- **arable weed**
  a group of wild plants associated with arable farmland. Some plants in this category were once so common that they were formerly regarded as agricultural pests but have declined dramatically with the onset of agricultural intensification. Examples of once-common species that are now scarce include the cornflower (Centaurea cyanus) and certain species of poppies (Papaver spp.).

- **arboreal**
  tree- dwelling or connected with trees.

- **arboriculture**
  the cultivation and management of individual trees. (Compare with 'Silviculture').

- **arthropod**
  a highly diverse group of invertebrate animals all with jointed limbs. It includes insects (e.g. butterflies), crustaceans (e.g. crabs) and arachnids (e.g. spiders).

- **artificial breeding**
  the term used to describe the process of inseminating a female animal with semen for conception.
- **artificial colouring**
  a colouring containing any dye or pigment manufactured by a process of synthesis or other similar artifice, or a colouring which was manufactured by extracting naturally produced dyes or pigments from a plant or other material.
- **artificial flavouring**
  artificial flavours are restricted to an ingredient which was manufactured by a process of synthesis or similar process. The principal components of artificial flavours usually are esters, ketones, and aldehyde groups. These ingredients are declared in the ingredients statement as 'Artificial Flavours' without naming the individual components.
- **artificial insemination**
  placement of fresh or frozen male semen into the female mechanically without normal sexual contact.
- **artificial vagina**
  the device used by Livestock. Improvement to collect semen from artificial breeding bulls.
- **as fed**
  refers to feed as it is consumed by an animal, including moisture.
- **ascites**
  a collection of fluid within the abdominal cavity that may contain cancer cells.
- **aseptic**
  preventing or free from contamination by microorganisms.
- **aseptic packaging**
  technique for creating a shelf-stable container by placing a commercially sterile product into a commercially sterile container in a commercially sterile environment. The sealed container is designed to maintain product sterility until the seal is broken.
- **asexual reproduction**
  any form of reproduction in which new individuals are derived from a single parent.

aspartame
white, almost odorless, crystalline powder with an intensely sweet taste that is used as an artificial sweetener. It is approximately 180 times as sweet as sucrose and is used to enhance the flavor of cold or uncooked foods. Aspartame tends to lose its sweetness in the presence of heat, moisture, and alkaline media. Excessive use of aspartame should be avoided by patients with phenylketonuria (PKU) because the substance hydrolyses to form aspartylph-enylalanine.

aspergillosis
a fungal infection.

aspirin
a nonsteroidal anti-inflammatory drug (NSAID) used to treat fever, pain, and inflammation in the body.

assay
technique for measuring a biological response.

assemblage
a collection of plants and/or animals that is characteristic of a particular habitat or environment.

association
association consisting of and governed by representatives of milk producer groups.

asthma
respiratory disorder charac-ter-ised by recurring episodes of wheezing on expiration/inspiration due to constriction of the bronchi, coughing and viscous mucoid bronchial secretions. Episodes may be precipitated by inhalation of allergens or pollutants, infection, cold air, vigorous exercise or emotional stress.

astrocytomas
tumours that are composed of brain cells called astrocytes. The different kinds of astrocytomas are identified by the way the cancer cells look under a microscope.

atmospheric deposition
deposition of chemicals on the land from the air. Nitrates, for example, that occur in the atmosphere (either produced naturally in thunderstorms or released by industrial processes) may be washed out into the soil in rainwater or other forms of precipitation.

- **atrophy**
a defect or failure of nutrition or physiologic function manifested as a wasting away or reduction in the size of cell, tissue, organ or body part.

- **attention deficit disorder (add)**
syndrome affecting children, adolescents and rarely adults, characterised by learning and behavioral abilities. Symptoms may be mild or severe and are associated with functional deviation of the central nervous system without signs of major neurological or psychiatric disturbance.

- **attention deficit hyperactivity disorder (ADHD)**
a childhood mental disorder with onset before age 7 and involving inattention, impulsivity, and hyperactivity. See attention deficit disorder.

- **attenuated**
weakened, with reference to vaccines, made from pathogenic organisms that have been treated so as to render them avirulent.

- **auditing**
checking of book keeping.

- **autism**
developmental brain disorder that causes learning disabilities, personality disorders, difficulty in communicating with others, and in severe cases may cause aggressive or unusual behaviors. Autism is usually diagnosed in early childhood (before age 3) and is characterised by a marked unresponsiveness to others and the surrounding environment.

- **autoimmune disease**
  1. a disease in which the body produces antibodies against its own tissues.
  2. a condition in which the body recognises its own tissues as foreign and directs an immune response against them.

- **autoimmunity**
a condition in which the body mounts an immune response against one of its own organs or tissues.

- **autologous**
the transfer of material such as bone marrow or skin from a donor to him or herself. To make this possible,

donated material is removed and stored prior to the procedure, which will make the donation necessary.

- **autologous lymphocytes**
a person's own white blood cells. Lymphocytes have a number of roles in the immune system, including antibody production, attacking and destroying cancer cells and producing substances that kill cancer cells.

- **autologous tumour cells**
cancer cells from the patient's own tumour.

- **automatic detacher or automatic take-off**
a device for sensing the end of milk flow in the milking machine which shuts of the milking vacuum and releases the milking machine from the cow's udder.

- **autosome**
any chromosome other than a sex chromosome.

- **average daily gain**
the average amount of daily live weight increase as applied to farm animals.

- **avirulent**
unable to cause disease.

- **axilla**
the armpit.

- **axillary lymph nodes**
lymph nodes in the armpit that drain the lymph channels from the breast.

- **b cells**
white blood cells that develop from bone marrow and produce antibodies.

- **b lymphocytes (b-cells)**
a class of lymphocytes, released from the bone marrow, that produce antibodies.

- **b3 antigen**
a protein found on some tumour cells.

- **babcock test**
traditional method of measuring the butterfat content of milk, which may be used for calibrating modern electronic testing devices.

- **bacillus subtilis**
a bacterium commonly used as a host in recombi-

nant DNA experiments. Important because of its ability to secrete proteins.

- **bacillus thuringiensis (bt)**
naturally occurring soil bacterium that generates a protein toxic to a variety of lepidoptera, such as corn borers, but is harmless to people and animals.

- **backgrounding**
the preparation of young cattle for a feedlot, getting them accustomed to confinement facilities and feeds.

- **bacteria**
  1. minute single cell living creatures which can grow almost anywhere including in dirt, manure or plant material.
  2. living single-cell organisms. Water, wind, insects, plants, animals can carry bacteria and people and survive well on skin and clothes and in human hair. They also thrive in scabs, scars, the mouth, nose, throat, intestines, and room-temperature foods. Often bacteria are maligned as the causes of human and animal disease, but there are certain types, which are beneficial for all types of living matter.

- **bacterial toxin**
a poison made by a bacterium that can be modified to kill specific tumour cells without harming normal cells.

- **bactericide**
an agent or substance capable of destroying bacteria.

- **bacterin**
a suspension of killed or attenuated bacteria (vaccine) used to increase disease resistance.

- **bacteriophage**
virus that lives in and kills bacteria. Also called phage.

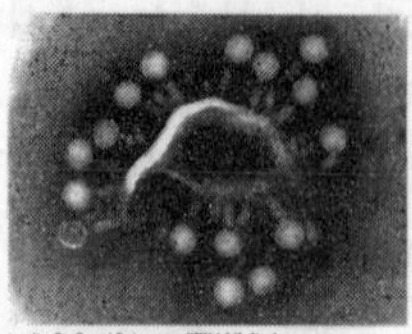

Phage des Bakteriums Acetobacter suboxydans

Quelle: Dr. Bernd Bohrmann, PBSV-NT, Roche

- **bacteriostatic**
describes a substance that prevents the growth of bacteria but does not kill them.

- **bacterium**
any of a large group of microscopic organisms with a very simple cell structure.

Some manufacture their own food, some live as parasites on other organisms, and some live on decaying matter.

- **balanced ration**
the daily food allowance of an animal, mixed to include suitable proportions of nutrients required for normal health, growth, production, and well-being. A ration containing all the dietary requirements to meet the purpose for which it is being fed.

- **baled hay**
forage that has been compressed into a bale (round or rectangular) to save space in storage and aid in handling.

- **balloon dilator**
an inflatable device designed to gently stretch or dilate mall tubular tracts. When the injected liquid is contrast or ye, the process can be viewed by X-ray. Most often used to pen the lower ureter for instrument passage, treat scarred areas of the ureter or prepare a passageway for percutaneous rocedures directly on the kidney.

- **barn cleaner**
usually a chain linked system of paddles that moved manure from gutters, up a chute, into a waiting manure spreader. Most often seen in tie-stall or stanchion barns.

- **barren**
sterile, infertile, nonbreeder, incapable, of producing offspring. Also, open or not pregnant.

- **barrow**
a male hog that has been castrated before reaching sexual maturity.

■ **basal metabolism (BM)**
the chemical changes that occur in the cells of an animal in the fasting or resting state when it uses just enough energy to maintain vital cellular activity, respiration and circulation as measured by the basal metabolic rate (BMR).

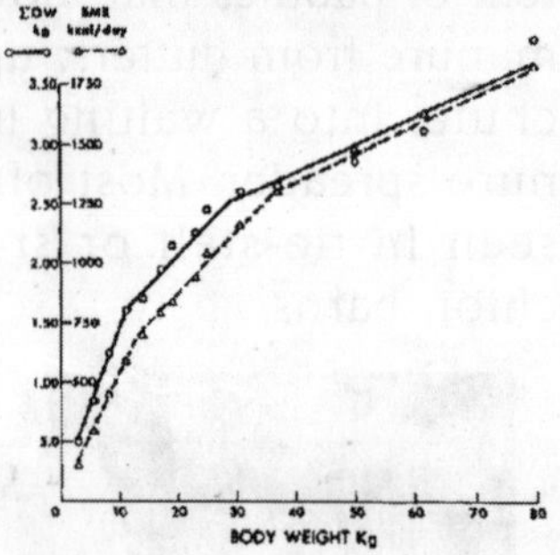

■ **basal rosette**
a term used in describing the form in which certain plants develop and grow. Their leaves emerge in a whorl from very close to the surface of the ground. This may be an adaptation to avoid the growing point being grazed by large animals.

■ **base**
1. an acid neutralising substance such as an antacid.
2. a key component of DNA and RNA molecules. Four different bases are found in DNA: adenine (A), cytosine (C), guanine (G) and thymine (T). In RNA, uracil (U) substitutes for thymine. Also known as nitrogenous bases. A base, a phosphate molecule and a sugar joined together constitute a nucleotide.
3. one of the four chemical units (nucleotides) arranged along the DNA or RNA molecule.

■ **base pair**
1. two nucleotide bases on different strands of the nucleic acid molecule that bond together. The bases can pair in only one way: adenine with thymine (DNA) or uracil (RNA) and guanine with cytosine.
2. a pair of nucleotides on complementary strands of DNA or RNA. Each nucleotide base can pair with only one of the three other bases, thereby determining the sequence of a complementary strand.

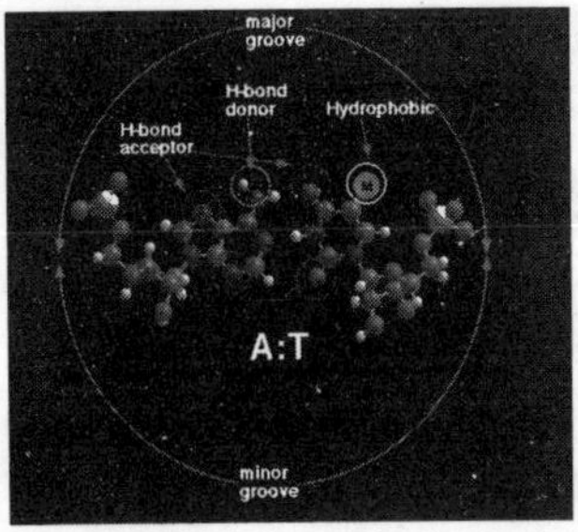

- **base-rich**
containing an abundance of basic (i.e. less acidic) materials. Soils derived from limestones, for example, tend to be base- rich and are more alkaline than acidic.

- **baste**
to moisten meat or other food while cooking. Melted butter or other fat, meat drippings, or liquid such as stock is spooned or brushed on food as it cooks to moisten it.

- **basted or self basted**
bone-in poultry products that are injected or marinated with a solution containing butter or other edible fat, broth, stock or water plus spices, flavour enhancers and other approved substances must be labelled as basted or self basted. The maximum added weight of approximately 3% solution before processing is included in the net weight on the label. Label must include a statement identifying the total quantity and common or usual name of all ingredients in the solution, e.g., 'Injected with approximately 3% of a solution of ____________ (list of ingredients).' Use of terms 'basted' or 'self-basted' on boneless poultry products is limited to 8% of the weight of the raw poultry before processing.

- **bay 12-9566**
an investigational drug that prevents the growth of new blood vessels to a tumour.

- **BCG vaccine**
the Bacille Calmett-Guerin (BCG) vaccine is a drug that activates the body's immune system.

- **bcl-2 antisense/g3139**
  an experimental drug that may kill cancer cells by blocking the production of a protein that makes cancer cells live longer.

- **bedded pack**
  open housing in a barn that is commonly used in conjunction with an outside feeding area.

- **bedding**
  material used to absorb moisture and provide cushion. It is easily cleaned to provide a clean, dry surface and reduce the incidence of mastitis. Possible bedding materials include: straw, sawdust, wood chips, sand, ground limestone, separated manure solids, shredded newspaper, corn stalks, bark, peanut hulls, sunflower hulls and rice hulls.

- **beef**
  meat from full-grown cattle about two years old. 'Baby beef' and 'calf' are interchangeable terms used to describe young cattle weighing about 700 pounds that have been raised mainly on milk and grass.

- **beef base - 15% beef or 10.5% cooked beef.**
  pork Base - 15% pork or 10.5% cooked pork.

- **beef patties**
  'beef patties' shall consist of chopped fresh and/or frozen beef with or without the addition of beef fat and/or seasonings.

- **beef suet**
  hard fat from kidneys and loin, mainly used for tallow. May be labelled as 'Beef Fat' or 'Beef Suet'.

- **beefy (beefiness)**
  a term used to designate the desirable physical conformation of a beef animal, as contrasted with a dairy animal which is trimmer (not beefy) and more angular.

- **beet sugar**
  means sugar, whether or not principally of crystalline structure, which is pro-

cessed directly or indirectly from domestically produced sugar beets (including sugar produced from sugar beet molasses).

- **benchtop formulation**

working at the laboratory scale on the lab bench, helping clients redesign their recipes (formulations) to work in an industrial context using industrial ingredients.

- **benign**

1. when used in a medical context, benign means non-cancerous.
2. not cancer; does not invade or spread to other parts of the body.

- **berne convention**

an international treaty signed by 40 nations. Formally it is called the 'Convention on the conservation of European wildlife and natural habitats. Bern/Berne 19.IX.1979.' Its stated aim is to 'conserve wild flora and fauna and their natural habitats'. The UK is a signatory to this treaty, which came into force in 1982. The Convention requires parties to give special attention to the conservation of species listed in the treaty's appendices numbered I, II and III.

- **biafine cream**

a topical preparation for the prevention and treatment of radiation-induced skin reactions.

- **bicarbonate**

a natural form of antacid or base that is normally excreted y the pancreas to neutralise the stomach acid as your food laves the stomach and enters the intestines. Bicarbonate is dissolved in the blood and converted to citrate by the kidneys. Common baking soda is chemically sodium bicarbonate.

- **bifidobacteria**

important bacterial culture naturally present in the human large intestine flora. Also used to make yogurt.

- **bio-**

prefix meaning 'biologically based', often attached to various verbs and nouns (e.g., bioprocessing, biofactory) in a biotechnology context.

■ **bioassay**
determination of the effectiveness of a compound by measuring its effect on animals, tissues or organisms in comparison with a standard preparation.

■ **bioaugmentation**
increasing the activity of bacteria that break down pollutants by adding more of their kind. A technique used in bioremediation.

■ **biocatalyst**
in bioprocessing, an enzyme that activates or speeds up a biochemical reaction.

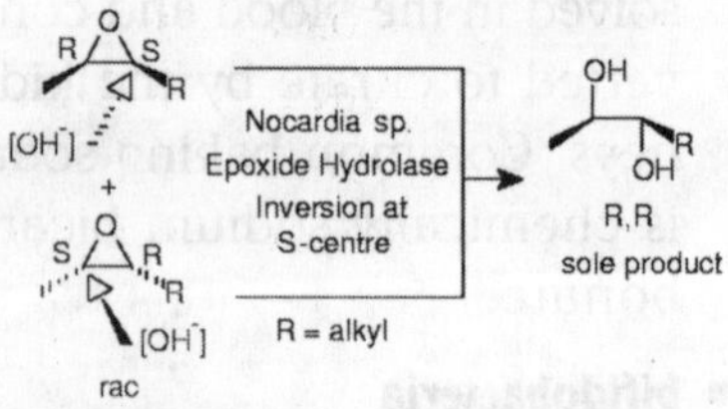

■ **biochemical**
the product of a chemical reaction in a living organism.

■ **biochip**
an electronic device that uses organic molecules to form a semiconductor.

■ **bioconversion**
chemical restructuring of raw materials by using a biocatalyst.

■ **biodegradable**
capable of being reduced to water and carbon dioxide by the action of microorganisms.

■ **biodiversity**
the variability among living organisms from all sources including terrestrial, marine and other aquatic ecosystems and the ecological complexes of which they are part; this includes diversity within species, between species and of ecosystems.

■ **biodiversity action plan**
a plan, programme or strategy for the conservation and sustainable use of biological diversity.

■ **biodiversity indicator**
indicators are quantified information that help to explain how things are changing over time. Biodiversity indicators are intended as one measure of how the natural world is responding to human- induced impacts.

For this reason they are able to gauge of the effectiveness of policies that aim at sustainable development.

- **bioenrichment**
  a bioremediation strategy that involves adding nutrients or oxygen, thereby bolstering the activity of microbes as they break down pollutants.

- **bioinformatics**
  the science of informatics as applied to biological research. Informatics is the management and analysis of data using advanced computing techniques. Bioinformatics is particularly important as an adjunct to genomics research, because of the large amount of complex data this research generates.

- **biolistic device**
  a device that shoots microscopic DNA-coated particles into target cells.

- **biologic (biologic response modifier)**
  a substance that alters the growth or functioning of a cell. Includes hormones and compounds that affect the nervous and immune systems.

- **biological oxygen demand (BOD)**
  the amount of oxygen used for growth by organisms in water that contains organic matter.

- **biological response modifier**
  a substance that can improve the body's natural immune response to disease. Cytokine therapy is a form of biological response modifier therapy.

- **biological therapy**
  treatment to stimulate or restore the ability of the patient's immune system to fight infection and disease. Also called immunotherapy or biological response modifier (BRM) therapy.

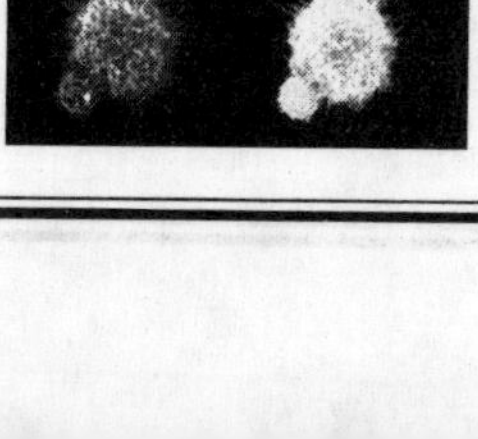

■ **biomarkers**
substances sometimes found in an increased amount in the blood, other body fluids, or tissues and which can be used to indicate the presence of some types of cancer.

■ **biomass**

1. the total weight of living matter in a population.
2. the totality of biological matter in a given area. As commonly used in biotechnology, refers to the use of cellulose, a renewable resource, for the production of chemicals that can be used to generate energy or as alternative feed stocks for the chemical industry to reduce dependence on nonrenewable fossil fuels.
3. materials and waste from living things that can be used as a source of fuel, energy, commodity chemicals, animal feed, and specialty products (e.g., flavours, fragrances, pigments).

■ **biomaterials**

1. biological molecules, such as proteins and complex sugars, used to make medical devices, including structural elements used in reconstructive surgery.
2. materials with novel chemical, physical, mechanical, or 'intelligent' properties, produced through processes that employ or mimic biological phenomena.

■ **biomimetic**
mimicking a biological substance or process.

■ **biomolecular materials**
complex biological macromolecules that can have unusual combinations of properties (e.g., strong and supple yet lightweight). These materials comprise a subset of biomaterials.

■ **bioprocess**

a process in which living cells, or components thereof, are used to produce a desired product.

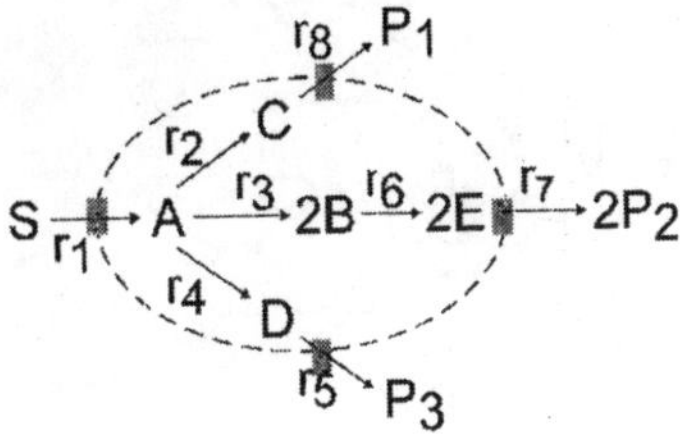

■ **biopsy**

the removal of cells or tissues, which are then examined under a microscope to check for cancer cells. When only a sample of tissue is removed, the procedure is called incisional biopsy or core biopsy. When the whole tumour is removed, it is called excisional biopsy. Removing tissue or fluid with a needle for microscopic examination is called needle biopsy or needle aspiration. Removing tissue from a lymph node is called lymph node biopsy.

■ **biopsy specimen**

tissue removed from the body and examined under a microscope to determine if disease is present.

■ **bioreactor**

1. vessel used for bioprocessing.
2. a chamber where biological growth, product formation and/or enzymatic reactions occur.

■ **bioremediation**

1. the use of microorganisms to remedy environmental problems, rendering hazardous wastes nonhazardous.
2. the use of living organisms or their products to degrade wastes into less toxic or non-toxic products and to concentrate and immobilise toxic elements, such as heavy metals. Bioremediation may be employed either on site or in bioreactors, using either native organisms or newly introduced ones, including genetically engineered varieties. The choice of organisms and treatments is determined by the nature of the toxic material and the biological and physical characteristics of the affected system.

- **biosynthesis**
  production of a chemical by a living organism.
- **biotechnology**
  agricultural biotechnology is a collection of scientific techniques, including genetic engineering, that are used to create, improve, or modify plants, animals and microorganisms. Using conventional techniques, such as selective breeding, scientists have been working to improve plants and animals for human benefit for hundreds of years. Modern techniques now enable scientists to move genes (and therefore desirable traits) in ways they could not before - and with greater ease and precision.
- **biotechnology**
  the use of biological processes to solve problems or make useful products.
- **bioterrorism**
  intentional use of biological agents or toxins to cause a public health emergency or to threaten the integrity of the food and agricultural system.
- **biotransformation**
  the use of enzymes in chemical synthesis to produce chemical compounds of a desired stereochemistry.

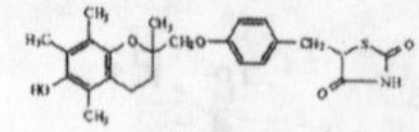

Troglitazone

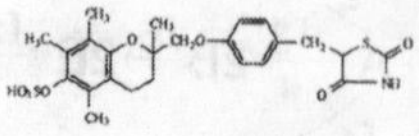

Metabolite 1 (Sulfate Conjugate)

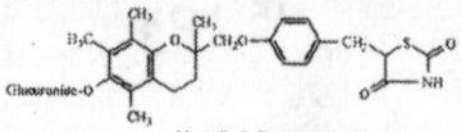

Metabolite 2 (Glucuronide Conjugate)

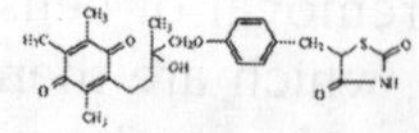

Metabolite 3 (Quinone Conjugate)

- **bispecific antibody**
  an antibody developed in the laboratory to recognise more than one protein on different cells. Some examples are bispecific antibodies 2B I, 52OC9xH22, MDX-H210, and MDX447.
- **blanket bog**
  a peatland formed in areas of high rainfall, covering large areas of flat and gently sloping ground.
- **blastocyst (blastula)**
  the 4- to 5-day-old ball of undifferentiated cells from which a prospective embryo develops. In mammals it consists of two distinct

parts: the inner cell mass and the trophoblast.

- **bleat**
  the vocal sound made by a goat.

- **blend price**
  the price paid producers for market milk when classified pricing is used. The blend price is an average of class prices weighted by the quantity of milk used in each class. See classified pricing.

- **bleomycin**
  a drug that inhibits cancer cell growth by interfering with the formation of DNA.

- **blind quarter**
  a quarter of an udder that does not secrete milk or one that has an obstruction in the teat that prevents the removal of milk. A non-functional mammary gland.

- **bloat**
  a disorder of ruminants usually characterised by an accumulation of gas in the rumen.

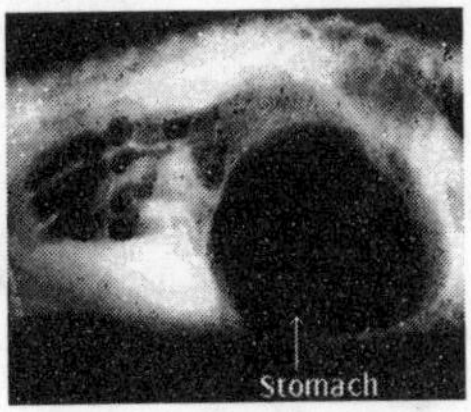

- **blood plasma**
  the liquid portion of blood in which the corpuscles of blood cells are suspended.

- **blood transfusion**
  the transfer of blood or blood products from one person to another.

- **bloom**
  a term commonly used to describe the beauty and freshness of a cow in early lactation. A dairy cow in bloom has a smooth hair coat and presents evidence of milking ability (dairy character).

- **bms-182751**
  a platinum compound used in chemotherapy.

- **board certified**
  formal acknowledgement by a medical or surgical specialty board that the practitioner has met all requirements for specialisation set by that board and has passed one or more rigorous examinations in that field. While not a guarantee of competence, it suggests a high level of knowledge in one particular specialty.

■ **BOD (biochemical oxygen demand)**
an indirect measure of the concentration of biodegradable substances in an aqueous solution. The degradation of organic matter uses up dissolved oxygen.

■ **bog**
a peat land mire that receives water only from precipitation (rain, snow, fog, etc.) And are as a result nutrient-poor.

■ **boiling point elevation**
one of the colligative properties. The boiling point of a solution is increased over that of water by the presence of solutes, and the extent of the increase is a function of both concentration and molecular weight.

■ **bolus**
1. a single dose of drug.
2. regurgitated food that has been chewed and is ready to be swallowed; a large pill for dosing animals.

■ **bolus infusion**
the introduction of a single dose of fluid into a vein or artery.

■ **bone marrow**
the soft sponge-like material inside some bones. Blood cells are formed in the bone marrow. Bone marrow used in cancer therapy may be autologous (the patient's own marrow saved earlier), allogeneic (marrow from someone else) or syngeneic (marrow from an identical twin).

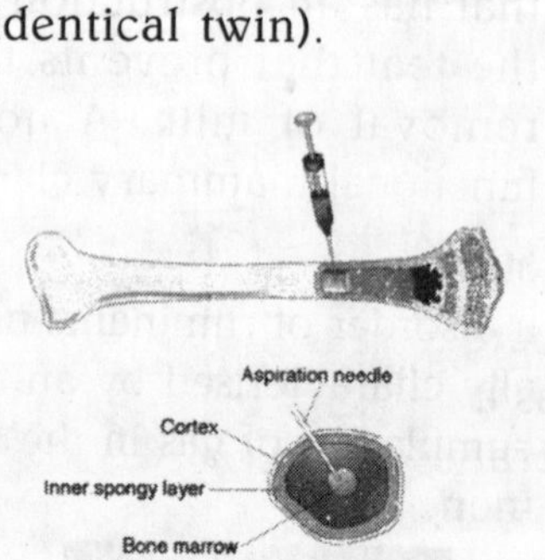

■ **bone marrow ablation**
destruction of cancerous bone marrow using radiation or drugs.

- **bone marrow metastases**
tumour cells that have spread from the original (primary) tumour and are growing in the bone marrow.

- **bone marrow transplantation**
a procedure in which doctors give marrow to replace marrow destroyed by treatment with high doses of anticancer drugs or radiation Transplantation may be autologous (the patient's own marrow saved earlier), allogeneic (marrow from someone else) or syngeneic (marrow from an identical twin).

- **bone meal**
animal bones that were steamed under pressure and then ground. It contains 1.5 to 2.5 percent nitrogen, 12 to 15 percent phosphorus, and 20 to 34 percent calcium. It is used as a fertiliser and as a mineral supplement for feeding farm animals.

- **bonn convention**
an international treaty formally called the 'Convention on the Conservation of Migratory Species of Wild Animals.' Its stated aim is the 'conservation of endangered and threatened species which undergo cyclic and predictable migration across one or more national boundaries'. The Convention requires parties to give special attention to the conservation of species in appendix I (where these occur in their territory) and to conclude agreements with other parties to protect other species (listed in appendix II) that are in 'unfavourable conservation status'.

- **boron neutron capture therapy**
a type of radiation therapy. The tumour cells give the patient an intravenous infusion containing boron, which is absorbed. Radiation is directed at the boron, killing the tumour cells and sparing the surrounding normal tissue.

■ **botulism**
a rare but serious paralytic illness caused by a nerve toxin that is produced by the bacterium Clostridium botulinum. There are three main kinds of botulism, one of which is food borne botulism caused by eating foods that contain the botulism toxin. Food borne botulism can be especially dangerous because eating a contaminated food can poison many people. All forms of botulism can be fatal and are considered medical emergencies. Good supportive care in a hospital is the mainstay of therapy for all forms of botulism.

■ **bovine**
pertaining to cattle, both beef and dairy.

■ **bovine growth hormone**
see BST or Somatotropin.

■ **bovine somatotropin (BST)**
a hormone secreted by the bovine pituitary gland. It is used to increase milk production by improving the feed efficiency in dairy cattle milk. Also called bovine growth hormone.

■ **bovine spongiform encephalopathy (BSE)**
commonly known as 'mad cow disease', BSE is a slowly progressive, incurable disease affecting the central nervous system of cattle, first diagnosed in the United Kingdom in 1986. BSE belongs to a family of diseases known as the transmissible spongiform encephalopathies (TSEs). Consumption by cattle of animal feed containing TSE-contaminated ruminant protein has been cited as one possible means of transmission. Scientific evidence supports a causal relationship between BSE outbreaks in Europe and more than 120 recent European cases of a human TSE, variant Creutzfeldt-Jakob disease (vCJD). TSE animal diseases are found in the United States, including scrapie in sheep and goats and chronic wasting disease in deer and elk.

- **brachytherapy**
internal radiation therapy using an implant of radioactive material placed directly into or near the tumour.

- **bracken**
a species of fern (Pteridium aquilinum) that forms large stands.

- **brackish**
slightly salty water such as that found in a river near its discharge point into the sea.

- **brain stem glioma**
tumours located in the part of the brain that connects to the spinal cord (the brain stem). They may grow rapidly or slowly, depending on the grade of the tumour.

- **brain stem tumour**
an abnormal growth on the back of the head where the spinal cord joins the brain.

- **bran**
the seed coat of wheat and other cereal grains which is separated from flour and used as animal food.

- **brca1 and brca2 (breast cancer genes 1 and 2)**
two genes that normally help to restrain cell growth, but which can contain certain genetic mutations associated with the development of breast and ovarian cancer. Note, however, that inherited BRCA1 and BRCA2 mutations are thought to account for less than 10 percent of all breast and ovarian cancers. Recent evidence suggests that somatic cell genetic mutations (i.e., noninherited genetic mutations) in these two genes may also play a role in the development of cancer.

- **breed**
  1. the type of animal eg Holstein Friesian (black and white), Jersey (brown), Ayrshire (red and white).
  2. animals having a common origin and characteristics that distinguish them from other groups within the same species.

- **breed average**
the average milk production of cows for a given dairy breed.

- **breeding services**
upgrading of animals through introduction of improved genes.

- **breeding value (genetic value)**
  the genetic ability or merit of an animal for a given trait, for example, secreting milk. One-half of this genetic ability is transmitted to offspring.

- **breeding worth**
  an index used to indicate the genetic potential of an animal to earn profit for its owner.

- **brine**
  to treat with or steep in brine. (Noun) A strong solution of water and salt, and a sweetener such as sugar, molasses, honey, or corn syrup may be added to the solution for flavour and to improve browning.

- **brine curing**
  brine curing (or wet curing) is the most popular way of producing hams. It is a wet cure whereby fresh meat is injected with a curing solution before cooking. Brining ingredients can be salt, sugar, sodium nitrite, sodium nitrate, sodium erythorbate, sodium phosphate, potassium chloride, water and flavourings. Smoke flavouring (liquid smoke) may also be injected with brine solution. Cooking may occur during this process.

- **brisket board**
  a raised part of the freestall platform about 6.5 feet in front of rear of the stall to keep cows positioned properly while lying. Usually made of wood or plastic, but occasionally concrete.

- **British breeds**
  those native to Great Britain, such as Hereford, Angus and Shorthorn.

- **broadleaved**
  any tree that is not a conifer.

- **broad-spectrum antibiotic**
  an antibiotic that is active against a large number of microbial species.

- **broiler**
  a young chicken produced for meat.

- **broiler or fryer**
a broiler or fryer is a young chicken, usually under 13 weeks of age, of either sex, that is tender-meated with soft, pliable, smooth-textured skin and flexible breastbone cartilage.

- **broken udder**
term used to describe an udder that is loosely attached or pendulous.

- **bronopol**
a noncorrosive milk preservative.

- **brood**
all of the offspring that hatch from a single clutch of eggs. To incubate eggs.

- **broodfish**
fish kept for egg production, including males. Broodfish produce the fertilised eggs that go to hatcheries.

- **browse**
small, woody plants with fedder shoots used as feed by goats, sheep and cattle.

- **broxuridine**
a drug that makes cancer cells more sensitive to radiation.

- **brucellosis**
contagious abortion in beef and dairy cattle, caused by Brucella abortus.

- **bryophytes**
the group of plants that includes mosses and liverworts. Members of the group do not have a 'vascular' system of vessels for conducting liquids (Compare with vascular plant). They are anchored to the substrate by 'rhizoids' that differ from true roots because they lack conducting vessels.

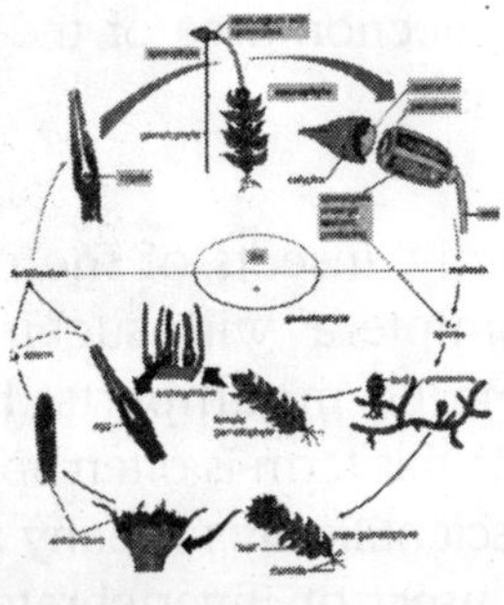

- **bryostatin-1**
a drug used for its antitumour activity.

- **BAT (bovine somatotropin)**
a complex, biologically active protein secreted by the

pituitary gland. It stimulates body cell growth and milk production available as a synthetically produced product for use in cattle. Sometimes called 'growth hormone' or BGH.

- **buck**
  a sexually mature male goat.

- **bucket milking system**
  a system in which the milk coming from the cow is drawn into a bucket or pail and manually transferred to a collection area or the milk house.

- **bug**
  strictly, insects of the order 'hemiptera' with sucking or piercing mouthparts. However, this term is often applied unscientifically to many types of insect or invertebrate.

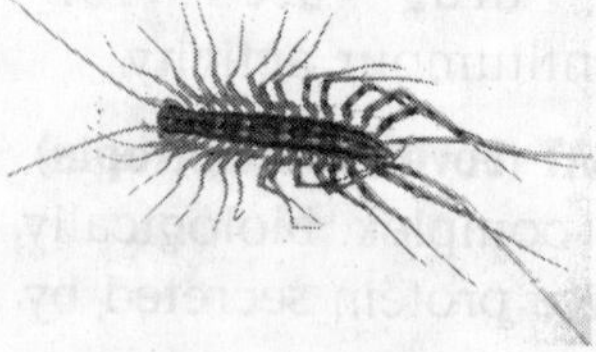

- **bulk milk**
  milk stored on the farm in a tank (or tanks), as compared with that stored in cans.

- **bulk milk tank**
  a refrigerated stainless steel vessel in which milk is cooled quickly to 2 to 4øC. (35 to 39øF.) And stored until picked up in a bulk tank truck.

- **bulk tank**
  a refrigerated, stainless steel vessel in which milk is cooled quickly to 2 to 4°C (35-39°F) and stored until collected by a bulk tank truck for shipping to the milk plant.

- **bull**
  a sexually mature, uncastrated bovine male, generally employed for breeding purposes.

■ **bulling**
a cow in heat (estrus).

■ **bunk**
a feed trough or feeding station for cattle.

■ **bunker**
sometimes called Bunker Silo. A flat rectangular structure with concrete floors and walls used to ensile and store forages.

■ **buserelin**
a drug used to block hormone production in the ovaries or testicles.

■ **buthionine sulfoximine**
an investigational drug that may help prevent resistance to some chemotherapy drugs.

■ **butter**
soft spread made by churning cream. Contains about 80% butterfat, which is partially crystallised.

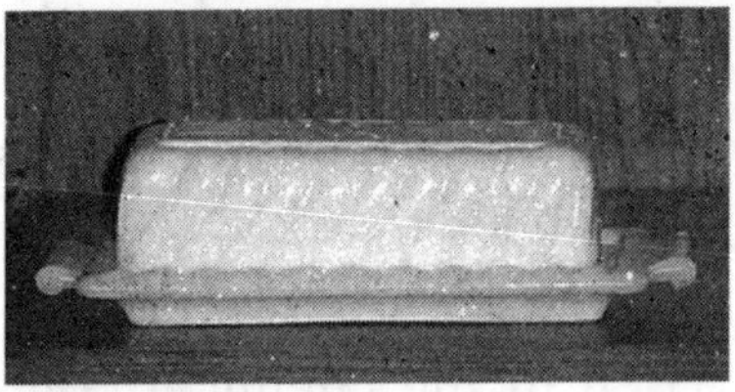

■ **buttermilk**
a form of cultured milk made from fresh low-fat or skim milk. Most of the milk fat is remove, and a special bacterial culture is added to product the desired acidity, body, flavour and aroma of buttermilk.

■ **buttermilks**
there are two kinds of buttermilk. Natural buttermilk is the by-product of churning cream into butter. Cultured buttermilk is usually made from skim milk by adding a starter culture and incubating until lactic acid develops to about 0.8 percent. Salt is usually added to accentuate the flavour.

■ **CA**
the abbreviation for the combination chemotherapy drugs cyclophosphamide and doxorubicin.

■ **CAF**
an abbreviation for the combination chemotherapy

drugs cyclophospharnide, doxorubicin, and fluorouracil. This combination is also known as FAC.

- **calcareous**

  rich in the chemical compound calcium carbonate ($CaCO_3$). For example, calcareous grassland develops on soils derived from calcium carbonate- rich rocks such as limestone.

- **calcitonin**

  a hormone made in the thyroid gland that increases calcium deposition in bone and increases urinary calcium. The opposite function to parathyroid hormone.

- **calcitriol**

  1. another name for activated Vitamin D, also called Vitamin D3.
  2. a compound made in the lab that is chemically similar to vitamin D.

- **calcium**

  1. a mineral naturally present in milk. Important in humans for strong teeth and healthy bones.
  2. an important element that helps make bone and teeth hard. It's necessary for proper functioning of muscles and nerves. Normally regulated in the body primarily by the parathyroid glands. In the diet, it can be found mainly in dairy products like milk and cheese.
  3. a mineral found in teeth, bones, and other body tissues.
  4. an essential nutrient for the growth and maintenance of healthy bones and teeth. It is widely found in dairy products.

- **calcium carbonate**

  a form of calcium used as a dietary supplement.

- **calcium oxalate**

  a compound made from calcium and oxalate. It can be in the monohydrate form, which is extremely hard, or dehydrate which is quite brittle.

- **calcium phosphate**

  another compound that forms kidney stones. Calcium phosphate is more likely to occur due to a medical problem or illness. It is formed most often in alkaline solutions. There is a danger in treating some patients with uric acid stones where too much citrate or bicarbonate can alkalinise the urine too severely and allow calcium phosphate, which will not dissolve, to coat or cover the uric acid.

- **calculi**
  the plural of calculus. Refers to many stones.

- **calculus**
  another name for a kidney stone. Actually, Latin for pebble or stone.

- **calf**
  a young male or female bovine. Usually referred to as calves until reaching sexual. maturity.

- **callus**
  a cluster of undifferentiated plant cells that can, in some species, be induced to form the whole plant.

- **calving**
  the term used when cows give birth.

- **calving interval**
  the average number of days between the latest two calving dates for individual cows or a herd.

- **camptothecin**
  belongs to the group of anticancer drugs called topoisomerase inhibitors.

- **camptothecin analogue**
  an anticancer drug related in structure to camptothecin, a topoisomerase inhibitor. One such drug is aminocamptothecin.

- **campylobacter**
  campylobacter is a bacterium that is commonly found in the intestinal tracts of cats, dogs, poultry, cattle, swine, rodents, monkeys, wild birds, and some humans. The bacteria pass through feces to cycle through the environment and are also in untreated water. Campylobacter jejuni (C. jejuni), the strain associated with most reported human infections, may be present in the body without causing illness.

- **campylobacteriosis**
  a diarrheal disease often caused by the type of bacteria known as Campylobacter jejuni (C. jejuni) associated with poultry, raw milk, and water. There are an estimated 2.5 million

cases annually in the United States with 200 to 730 deaths. Campylobacteriosis has been linked to Guillain-Barre syndrome (a disease which paralyses limbs and breathing muscles) as well as Epstein-Barr, Cytomegalovirus and other viruses.

- **can**
a receptacle generally having less than 10 gallon capacity (consumer or institutional sizes), also means to pack a product in a can or a wide-mouth glass container for processing, shipping or storage.

- **can milk**
that milk placed in 10-gallon cans on the farm for cooling and transportation to a dairy product manufacturing plant.

- **canalisation**
engineering to constrain rivers or streams to a particular course.

- **candidiasis**
any infection caused by a species of Candida, especially Candida albicans, characterised by pruritus, peeling, and easy bleeding. Diaper rash, intertrigo, vaginitis, and thrush are common topical manifestations of candidiasis. See **yeast-free diet**.

- **cane sugar refiner**
means a person who processes raw cane sugar into direct- consumption sugar. The same person may be both a 'cane sugar refiner' and either a 'sugarcane processor' or 'sugar beet processor' or both.

- **cane syrup**
means concentrated cane juice from which no sucrose has been extracted.

- **cannula**
a special tube designed for placing drugs into the udder through the teat end and streak canal.

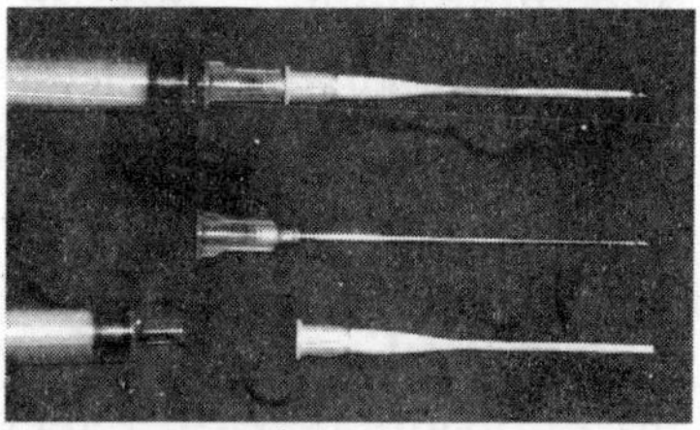

- **canopy**
the branches, leaves etc. formed by woody plants (trees, shrubs) some way above the ground.

- **capacity**
when applied to the udder refers to the cow's ability to produce and store milk.

- **capacolla**
boneless pork shoulder butts which are dry cured, not necessarily cooked.

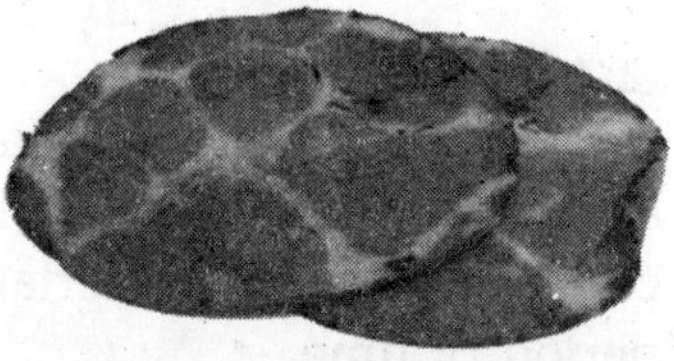

- **capon**
a surgically unsexed male chicken, usually under eight months of age, that is tender-meated with soft, pliable, smooth-textured skin.

- **capriculture**
the study of goat keeping.

- **captive bolt**
an instrument used to stun cattle prior to slaughter. The bolt is driven into the animal's brain, rendering it unconscious.

- **carbogen**
an inhalant of oxygen and carbon dioxide that increases the sensitivity of tumour cells to the effects of radiation therapy.

- **carbohydrate**
a type of biological molecule composed of simple sugars such as glucose. Common examples include starch and cellulose.

- **carboniferous**
the geological era that spanned the period 345 - 280 million years before the present. It is so-named because it is the time during which much of Britain's coal measures were laid down.

- **carboplatin**
an alkylating-like chemotherapy agent that contains platinum in its structure. Also called Paraplatin.

- **carboxyamidotriazole**
belongs to a group of chemotherapy drugs that affect

how cells use calcium. Also known as carboxyaminoimidazole or CAI.

- **carcass**
  all parts of any slaughtered livestock.

- **carcinoembryonic peptide-1**
  a protein that can stimulate an immune response to certain tumours. Also known as CAP-1

- **carcinogen**
  cancer causing agent.

- **carcinoma**
  a cancer of the tissue that covers the internal and external surfaces of the body.

- **cardiopulmonary**
  related to the heart and lungs.

- **carnivorous**
  meat- eating.

- **carob**
  an eastern Mediterranean evergreen tree (Ceratonia siliqua) in the pea family, having pinnately compound leaves and large, dark, leathery pods. The pod of this plant, containing a sweet edible pulp and seeds that yield a gum used as a stabiliser in food products. Also called algarroba. An edible powder or flour made from the ground seeds and pods of this plant, often used as a substitute for chocolate.

- **carotenoid**
  a substance found in yellow/orange fruits and vegetables and dark green leafy vegetables that may prevent the development of cancer.

- **carrion**
  dead and rotting carcasses of animals.

- **carrying capacity**
  the number of animals that a pasture can properly carry with feed for a certain period of time.

- **casei**
  a type of bacterial culture, Lactobacillus casei, added to YoPlus yogurts for gut health benefits.

- **casein**
  1. major protein in cow's milk, about 80% of total pro-

tein content. Main component of cheddar cheese.
2. a protein found in milk and which is the main ingredient of cheese.
3. a white, tasteless, odorless protein precipitated from milk by rennin. It is the basis of cheese and is used to make plastics, adhesives, paints and foods. See dairy.
4. the major protein of milk and cheese.

- **casein micelle aggregation**
  caseins are able to aggregate if the surface of the micelle is reactive. The Schmidt model further illustrates this.

- **Casein micelle aggregation**
  although the casein micelle is fairly stable, there are four major ways in which aggregation can be induced:
  1. chymosin - rennet or other proteolytic enzymes as in Cheese manufacturing.
  2. acid.
  3. heat.
  4. age gelation.

- **caseins**
  the casein content of milk represents about 80% of milk proteins. The principal casein fractions are alpha(s1) and alpha(s2)-caseins, ß -casein and kappa-casein. The distinguishing property of all caseins is their low solubility at pH 4.6. The common compositional factor is that caseins are conjugated proteins, most with phosphate group(s) esterified to serine residues. These phosphate groups are important to the structure of the casein micelle. Calcium binding by the individual caseins is proportional to the phosphate content.
  The conformation of caseins is much like that of denatured globular proteins. The high number of proline residues in caseins causes particular bending of the protein chain and inhibits the formation of close-packed, ordered secondary structures. Caseins contain no disulfide bonds. As well, the lack of tertiary structure accounts for the stability of caseins against heat denaturation because there is very little structure to unfold. Without a tertiary structure there is consider-

able exposure of hydrophobic residues. This results in strong association reactions of the caseins and renders them insoluble in water.

- **casing**
a membranous case for processed meat.

- **castration**
1. an operation which removes the testicles from an 'entire' male animal.
2. the removal of male reproductive organs (testicles) to reduce sexual activity and increase fattening of meat animals.

- **catalyst**
an agent (such as an enzyme or a metallic complex) that facilitates a reaction but is not itself changed during the reaction.

- **caterpillar**
the larval stage of butterflies and moths.

- **catheter**
1. a long, slender tube that enters the body. Most often it's used to drain the bladder but a smaller version is used for special X-rays called Retrograde Pyelograms. Catheters are flexible and usually made of latex or silicone.
2. a flexible tube used to administer fluids into or withdraw fluids from the body.

- **CD34 antigen**
a protein found on the surface of some bone marrow and blood cells.

- **CEF**
an abbreviation for the combination chemotherapy drugs cyclophosphamide, epirubicin, and fluorouracil. This combination is also known as FEC,

- **celecoxib**
a nonsteroidal anti-inflammatory drug.

- **celiac disease (celiac sprue)**
an inborn error of metabolism characterised by the inability to hydrolyse peptides contained in gluten. The disease affects adults and young children, who suffer from abdominal distention, vomiting, diarra-hoea, muscle wasting and extreme lethargy. There may be secondary lactose intolerance. Most patients respond well to a high-protein, high-calorie, gluten-free diet.

- **cell**

1. all living organisms are composed of membrane-bound units known as cells. micro-organisms may be composed of single cells whereas there are a million million cells in a human being.
2. the smallest structural unit of a living organism that can grow and reproduce independently.
3. the smallest structural and functional unit of living organisms, generally having a nucleus, cytoplasm, and a surrounding membrane.

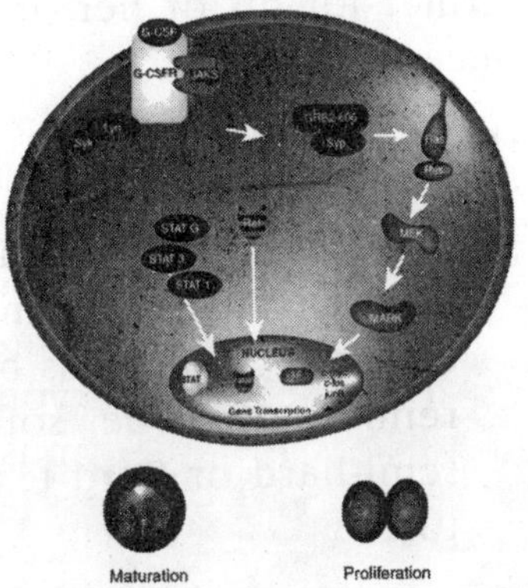

- **cell culture**

growth of cells under laboratory conditions.

- **cell line**

cells that grow and replicate continuously outside the living organism.

- **cell-mediated immunity**

acquired immunity in which T lymphocytes play a predominant role. Development of the thymus in early life is critical to the proper development and functioning of cell-mediated immunity.

- **cells**

a single unit of a body surrounded by a membrane, containing DNA and other small organs that help the cell function.

- **cellulose**

the principal carbohydrate constituent of plant cell membranes. It is made available to ruminants through the action of cellulolytic bacteria in the rumen.

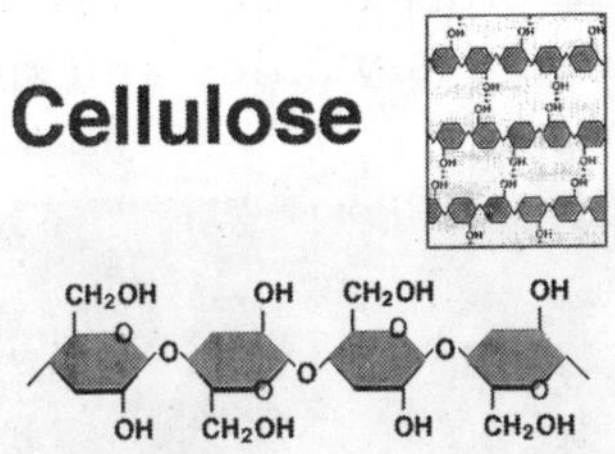

- **central venous access catheter**

a tube surgically placed into a blood vessel for continuous or repeated drug infusions. This device avoids the need for separate needle insertions for each infusion.

- **centre status**
  the state of health that a bull must be in to be able to be used for artificial breeding.
- **centrifugal separation**
  removal of one substance from another mechanically by centrifugal force, as with cream which is less dense than skim milk.
- **cereal**
  grass that produces an edible grain (e.g. oats, wheat, barley).
- **cervix**
  part of the reproductive organ of a cow.
- **cetacean**
  the order of mammals that includes whales and dolphins.

- **cgp 48664**
  an investigational chemotherapy drug that inhibits the growth of some tumours.
- **chain- harrowing**
  harrowing with chains drawn by a tractor.
- **chairperson**
  leader of the group with specific tasks.
- **chalk sticks**
  used to mark treated, fresh or special-needs cows.
- **characteristics**
  things that make something unique.
- **check-offs**
  the practice of deducting a portion of the payment to a farmer for his or her produce.
- **cheese**
  1. milk product.
  2. milk concentrate made by coagulation of fat and casein protein curds by rennet. May be soft, semi-hard or hard texture.
- **cheese curd**
  the clumps of casein and other milk components that are formed during the

cheese making process. These curds are then pressed into blocks or barrels for proper aging and curing of the cheese.

- **chemical genomics**
  using structural and functional genomic information about biological molecules, especially proteins, to identify useful small molecules and alter their structure to improve their efficacy.

- **chemical preservative**
  any chemical that, when added to a meat or meat food product, tends to prevent or retard deterioration thereof, but does not include common salt, sugars, vinegars, spices, or oils extracted from spices or substances added to meat and meat food products by exposure to wood smoke.

- **chemoprevention**
  the use of drugs, vitamins, or other agents to try to prevent or delay the development or recurrence of cancer. Chemoprevention drugs include celecoxib and fenretinide.

- **chemoprotective**
  agents that protect healthy tissue from the toxic effects of chemotherapy drugs. Chemoprotective drugs include deoxycytidine, fenretinide, pentetic acid calcium, and SR-29142.

- **chemosensitivity assay**
  a test to determine the most beneficial chemotherapy treatment by analysing the responsiveness of a tumour to a specific drug.

- **chemosensitiser**
  a drug which makes tumour cells more sensitive to the effects of chemotherapy.

- **chemotherapy**
  treatment with anticancer drugs. Chemotherapy may be taken by mouth or it may be put into the body by a needle inserted into a vein or muscle.

- **cheqmate nominated semen**
  the brand name for nominated semen supplied by Livestock Improvement.

- **chevon (fr)**
  goat meat used for food.

- **chick**
  a young bird that has recently hatched from the egg.

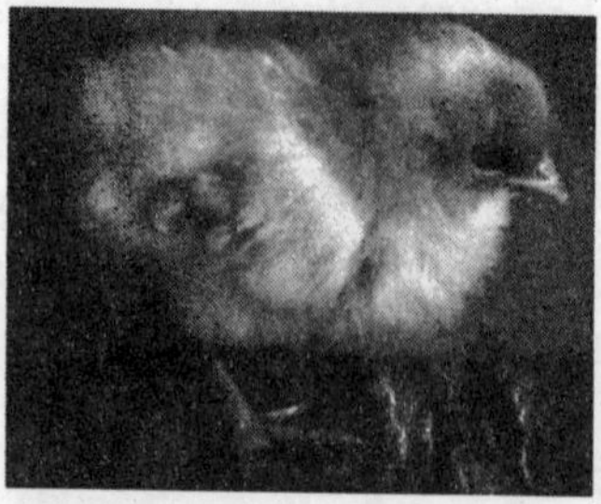

- **chimera**
  the individual (animal or lower organism) produced by grafting an embryonic part of one individual onto an embryo of either the same or a different species.

- **chlorambucil**
  a drug used to inhibit or prevent the development of new or abnormal tissue growth.

- **cholesterol**
  1. fat-like substance found in all animals, needed for the synthesis of vitamin D and hormones.
  2. a white, fat-soluble substance found in animal fats and oils, in bile, blood, brain tissue nervous tissue, the liver, kidneys, and adrenal glands. It is important in metabolism and is a precursor of certain hormones.

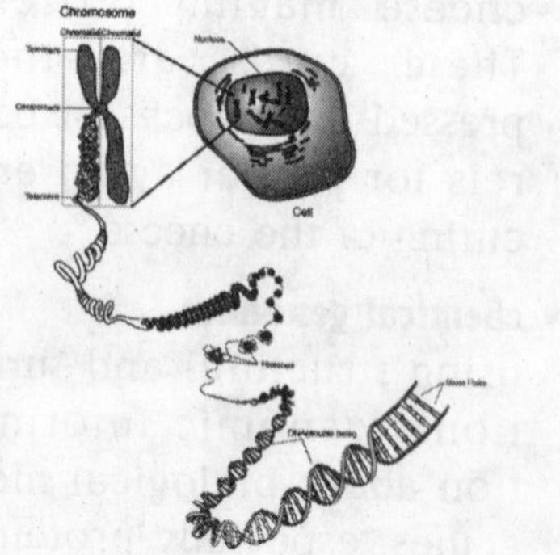

- **chromosome**
  1. the structures within cells that carry the genetic material.
  2. threadlike components in the cell that contain DNA and proteins. Genes are carried on the chromosomes.
  3. rod-like structures composed of highly coiled DNA and proteins.

- **chronic**
  1. a disease or condition which persists or progresses over a long period of time.
  2. used to describe recurring symptoms or disease.

- **chronic lymphoblastic lymphoma**
  a slowly progressing disease in which too many immature white blood cells called

lymphoblasts are found in the body.

- **chronic myclogenous leukemia**
  cancer in which too many white blood cells are made in the bone marrow. Also called chronic myeloid leukemia.

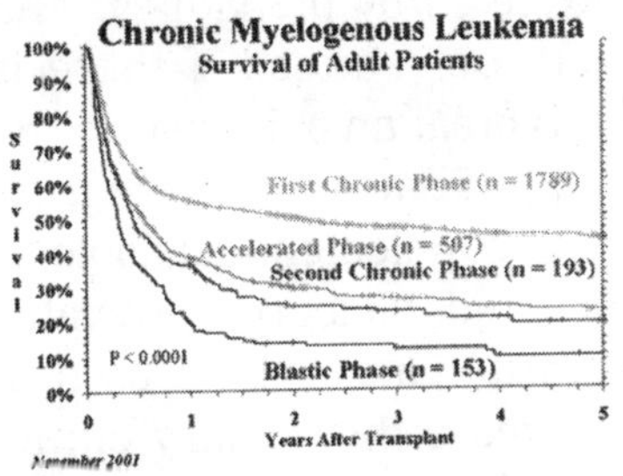

- **chronic phase chronic myelogenous leukemia**
  a phase that may last from several months to several years. Although there may be no symptoms of leukemia, there are few bone marrow cells, or blast cells, in the blood and bone marrow.

- **chronic wasting disease**
  chronic wasting disease (CWD) is a transmissible spongiform encephalopathy (TSE) of deer and elk. To date, this disease has been found only in cervids (members of the deer family) in North America. First recognised as a clinical 'wasting' syndrome in 1967 in mule deer in a wildlife research facility in northern Colorado, it was identified as a TSE in 1978. CWD is typified by chronic weight loss leading to death. There is no known relationship between CWD and any other TSE of animals or people.

- **chub**
  an acceptable name to denote a short, usually plump meat food product, unsliced in casing.

- **churn**
  container or machine in which butter is made from cream or milk by beating and shaking.

- **churning**
  1. agitation of whole milk or cream to make butter adding starters and rennet.
  2. the process of stirring and agitating cream in the process of making butter. Churning causes the fat globules in cream to clump together and separate from the liquid.

- **chymosin**
  a digestive enzyme contained in rennet that clots or curds milk. Also known as rennin.
- **ci-958**
  belongs to a group of chemotherapy drugs that affect how cells use DNA, the molecule that carries genetic information.
- **ci-980**
  a drug used in cancer chemotherapy.
- **cimetidine**
  a drug usually used to treat stomach ulcers. It is also used to control one type of white blood cell that controls immune responses. This drug is also known as Tagamet.
- **cip (cleaned-in-place)**
  most milk-handling equipment are cleaned by circulating washing solutions through them without disassembly.
- **cisplatin**
  a chemotherapy drug that contains platinum in its structure. Also called Platinol.
- **cites**
  convention on the International Trade in Endangered Species. The main international treaty that regulates the sale of exotic animal an plant species.
- **citrate**
  an important urinary acid neutraliser and inhibitor of kidney stone formation. It is excreted by the kidneys and will be reduced if there is too great an acid load in the body. Potassium citrate is an oral supplement available as a liquid, tablet or dissolvable crystal packet that's used to increase the urinary citrate level.
- **cladribine**
  a radio sensitisation drug that makes tumour cells more sensitive to the effects of radiation therapy.
- **clarification**
  the process of removing suspended material; in milk processing, clarification is achieved by centrifugal treatment which removes sediment (extraneous matter) and somatic cells.
- **classification**
  an appraisal program offered by a breed association to evaluate each animal's resemblance to the breed's

ideal. A numerical score is assigned to each animal.

- **clean**
  a lay term commonly used by cattle breeders to mean that a cow has shed her afterbirth. Also refers to the natural, wholesome flavour of dairy products.

- **climax**
  a plant community determined and maintained in a given area by the prevailing climatic and soil conditions essentially those which would develop and be present in the absence of human intervention.

- **clinical**
  symptoms are present, supportive therapy or treatment is necessary.

- **clinical studies**
  human studies that are designed to measure the efficacy of a new drug or biologic. Clinical studies routinely involve the use of a placebo group that is given an inactive substance that looks like the test product.

- **clinical trial**
  a research study to evaluate new methods of screening, prevention, diagnosis or treatment of cancer.

- **clint**
  flat top of a block of limestone pavement.

- **clodronate**
  a drug used as treatment for hypercalcemia (abnormally high levels of calcium in the blood) and for cancer that has spread to the bone (bone metastases), and may decrease pain, the risk of fractures and the development of new bone metastases.

- **clone**
  a term that is applied to genes, cells or entire organisms that are derived from and are genetically identical to a single common ancestor gene, cell or organism, respectively. Cloning of genes and cells to create many copies in the laboratory is a common procedure essential for biomedical research. Note that several processes commonly described as cell 'cloning' gives rise to cells that are almost but not completely genetically identical to the ancestor cell. Cloning of or-

ganisms from embryonic cells occurs naturally in nature (e.g., identical twins). Researchers have achieved laboratory cloning using genetic material from adult animals of several species including mice, pigs and sheep.

- **cloning**
  the use of techniques of molecular biology to produce multiple copies of segments of DNA, usually genes.

Cloning into a Yeast Artificial Chromosome (YAC)

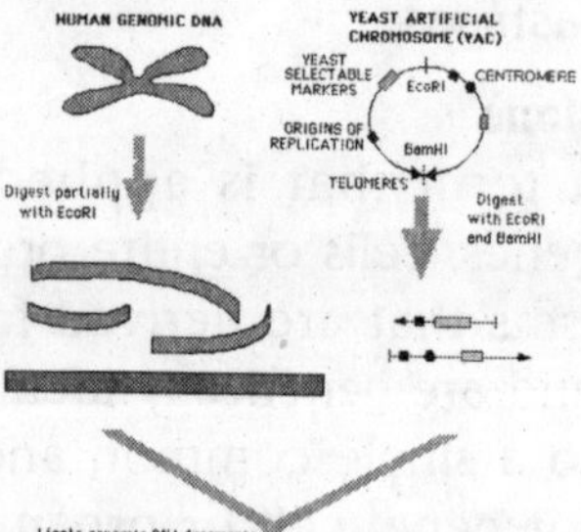

- **close breeding**
  a form of inbreeding, such as mating brothers to sisters, sire to daughter, and son to dam.

- **clostridium botulinum**
  the name of a group of bacteria commonly found in soil. These rod-shaped organisms grow best in low oxygen conditions. The bacteria form spores, which allow them to survive in a dormant state until exposed to conditions that can support their growth. Clostridium botulinum is the bacterium that produces the nerve toxin that causes botulism.

- **clot on boiling test**
  milk quality test through heating.

- **clotting**
  change from liquid milk to (semi-) solid by.

- **CMF**
  the abbreviation for the combination chemotherapy drugs cyclophosphamide, methotrexate, and fluorouracil.

- **CMFPT**
  the abbreviation for the combination chemotherapy drugs cyclophosphamide, methotrexate, fluorouracil, prednisone and tamoxifen.

- **CMP**
  the abbreviation for the combination chemotherapy drugs cyclophosphamide, mitoxantrone, and cisplatin.

- **CNS**
  the central nervous system (CNS) consists of the brain, spinal cord, and meninges (the surrounding membranes).

- **CNS metastases**
  central nervous system (CNS) metastases are tumour cells that have spread from the original (primary) tumour and are growing in the central nervous system.

- **CNS tumours**
  tumours of the brain and central nervous system, including brain stem glioma, craniopharyngioma, medulloblastoma, and meningioma.

- **coactivated t cells**
  cells that have been coated with monoclonal antibodies to enhance their ability to kill tumour cells.

- **coagulation**
  the act of changing from liquid form into a thickened mass.

- **coagulum**
  substance that changes from its liquid form into a thickened mass.

- **cock or rooster**
  a mature male chicken with coarse skin, toughened and darkened meat and hardened breastbone tip.

- **COD (chemical oxygen demand)**
  an indirect measure of the biochemical load exerted on the oxygen of a body of water when organic wastes are introduced into the water. If wastes are readily biodegradable, COD and BOD are nearly the same.

- **code date**
  found on all perishable products. Last date a product can be sold, it can usually be consumed for at least a week after that time, provided it has been cared for properly.

- **codex alimentarius commission**
  a joint commission of the Food and Agriculture Organisation (FAO) and the World Health Organisation, comprised of some 146 member countries, created in 1962 to ensure consumer food safety, establish fair practices in food trade, and promote the development

of international food standards. The Commission drafts nonbinding standards for food additives, veterinary drugs, pesticide residues, and other substances that affect consumer food safety. It publishes these standards in a listing called the 'Codex Alimentarius'.

- **codon**
a sequence of three nucleotide bases that specifies an amino acid or represents a signal to stop or start a function.

- **co-enzyme**
an organic compound that is necessary for the functioning of an enzyme. Co-enzymes are smaller than the enzymes themselves and sometimes separable from them.

- **co-factor**
a non-protein substance required for certain enzymes to function. Co-factors can be co-enzymes or metallic icons.

- **col-3**
an investigational drug that may stop tumour growth by preventing the development of blood vessels.

- **colic or renal colic**
the extremely severe pain associated with a kidney stone attack.

- **coliform**
any of various bacteria found in colon of vertebrates and excreted in the feces, coliform bacteria from human and animal feces may contaminate water and cause disease.

- **coliform bacteria**
bacteria from the intestinal tract of warm-blooded animals. Presence is considered indicative of fecal contamination.

- **colitis**
an inflammatory condition of the large intestine. Severe diarrhoea, bleeding and ulceration of the mucosa of the intestine characterise inflammatory bowel disease. Weight loss and pain are significant. Steroids, fluids, electrolytes, antibiotics, and careful attention to diet are usual modes of therapy. See **crohn's disease**.

- **collaboration matrix**
tool to assess possible collaboration with other organisations.

- **collecting system**

  the hollow, branched, interior portion of the kidney system that normally is full of urine. This is the target area for percutaneous procedures and the space in which most stones form. The exit of the collecting system (see **ueteropelvic junction**) is shaped like a funnel and leads into the ureter, the tube that connects the kidney with the urinary bladder.

- **collection barn**

  the barn where bulls are taken for semen collection.

- **collection centre**

  a place to collect milk in bulk.

- **colligative properties**

  properties which depend on the number of molecules in solution, a function of concentration and molecular weight, rather than just on the total percent concentration. Such properties include boiling point elevation, freezing point depression, and osmotic concentration.

- **colloidal calcium phosphate (CCP)**

  acts as a cement between the hundreds or even thousands of sub micelles that form the casein micelle. Binding may be covalent or electrostatic. Sub micelles rich in kappa-casein occupy a surface position, whereas those with less are buried in the interior. The resulting hairy layer, at least 7 nm thick, acts to prohibit further aggregation of sub micelles by steric repulsion. The casein micelles are not static; there are three dynamic equilibria between the micelle and its surroundings:

  1. free casein molecules and sub micelles.
  2. the free sub micelles and micelles.
  3. the dissolved colloidal calcium and phosphate.

- **colony**

  a group of animals or plants living together and dependent on each other to a greater or lesser extent (e.g. a colony of ants).

- **colony-stimulating factors (CSFS)**

  1. substances that stimulate the production of blood cells. Treatment with colony-stimulating factors (CSFS) can help in recovery from the effects of chemotherapy and

radiation therapy. Colony-stimulating factors include granulocyte colony-stimulating factors (G-CSF, filgrastim, Neupogen) and granulocyte-macrophage colony-stimulating factors (GM-CSF, sargramostim, Leukine, Prokine).
2. a group of lymphokines that induce the maturation and proliferation of white blood cells from the primitive cell types present in bone marrow.

- **colostrum**
  1. the first milk that a cow produces for 4-5 days after calving. It differs in composition and properties from regular milk and is high in antibodies.
  2. first milk following calving. High in fat, protein, and immunoglobulins that may be directly absorbed by the newborn calf in its first 24 hours of life.
  3. the first milk secreted after calving. Higher in total solids than normal milk.

- **colour ring**
a special ring attached to a bird's foot for the purposes of a particular study. These rings differ from the normal metal rings that record information for the purposes of tracking bird migration etc.

- **combination chemotherapy**
chemotherapy using more than one drug.

- **combinatorial chemistry**
a product discovery technique that uses robotics and parallel synthesis to generate and screen quickly as many as several million molecules with similar structure in order to find chemical molecules with desired properties.

- **co-metabolism**
a microbe oxidising not only its main energy source but also another organic compound.

- **commercial herd**
a herd of cows which is milked, the sale of the milk produces income for its owner.

- **commodity storage**
usually a steel framed shed that provides storage for commodity feeds, such as cottonseed, brewers grains, chopped hay etc.

- **compaction**
closely packed feed in the stomach and intestines of an animal causing constipation and/or digestive disturbances.

- **compartment**
a sub- division of a site for management purposes usually applied to parts of a nature reserve recognised in the reserve's management plan.

- **complementarity**
the relationship of the nucleotide bases on two different strands of DNA or RNA. When the bases are paired properly (adenine with thymine [DNA] or uracil [RNA], guanine with cytosine), the strands are complementary.

- **complementary DNA (cDNA)**
DNA synthesised from a messenger RNA rather than from a DNA template. This type of DNA is used for cloning or as a DNA probes for locating specific genes in DNA hybridisation studies.

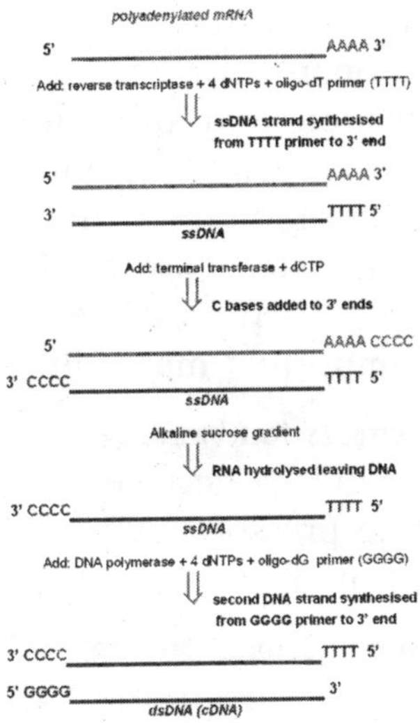

- **complete ration**
a blend of all feedstuffs (forages and grains) in one feed. A complete ration fits well into mechanised feeding and the use of computers to formulate least-cost rations. Sometimes called total mixed ration or TMR.

- **complete remission**
the disappearance of all signs of tumour.

- **complete response**
the disappearance of all signs of tumour.

- **components**
the nutritional solids in milk other than water, like fat, protein, lactose and minerals.

▪ **composting pad**
a concrete or hard packed surface that provides an area on which manure and discarded feed may be composted with ready access to aerate the composting materials.

▪ **compression bandage**
a bandage designed to provide pressure to a particular area.

▪ **computational biology**
a sub discipline within bioinformatics concerned with computation-based research devoted to understanding basic biological processes.

▪ **concentrate**
a feed high in nitrogen-free-extract (NFE) and total digestible nutrients (TDN) and low in crude fibre (less than 18 percent). Included are cereal grains, soybean oil meal, cottonseed meal, and by-products of the milling industry such as corn gluten and wheat bran. A concentrate may be low or rich in protein.

▪ **concentrated milks**
milk, skim milk and components of thes-e that have been concentrated by removing varying amounts of water under carefully controlled conditions of heat and/or vacuum. Concentrated milks may be reconstituted by adding appropriate quantities of water.

▪ **concentrates**
high energy or high protein feeds consisting primarily of the seed of the plant, but with out stems and leaves.

▪ **conception**
1. when a sperm and ovum combine to produce a pregnancy.
2. fertilisation or penetration of the ovum by a sperm cell.

▪ **conception rate**
total number of conceptions obtained divided by total number of services.

▪ **condition**
the amount of flesh (body weight), quality of hair coat, and general health of animals. Also called body condition.

▪ **confidence range (cr)**
indicates the accuracy in the estimation of a sire's genetic merit in a sire proof by giv-

ing a probable range for future summaries.

- **confinement**
  livestock kept in corrals or housing for maximum year round production. Facilities may be partial or complete, usually with a solid floor and enclosed or covered.

- **conformation**
  physical shape of an animal.

- **conformation**
  1. the body form or physical traits of an animal or parts of the animal in the case of udder conformation.
  2. the body form or physical traits of an animal; its shape and arrangement of parts. Conjunctiva The tissue covering the anterior portion of an eyeball.

- **congestive heart failure**
  weakness of the heart muscle, usually due to heart disease but sometimes due to other conditions, that causes a buildup of fluid in body tissues.

- **conifer**
  cone- bearing trees or shrubs that are mostly evergreen. There are only three native conifers in Britain: Scots Pine (Pinus sylvestris); Yew (Taxus baccata) and Juniper (Juniperus communis). Many non- native conifers, however, have been planted in this country for commercial purposes.

- **conjugation**
  sexual reproduction of bacterial cells in which there is a one-way exchange of genetic material between the cells in contact.

- **consolidation therapy**
  a second round of chemotherapy to further reduce the number of cancer cells.

- **constant effort (bird ringing) site**
  a site at which birds are caught and ringed year after year in order to study patterns in numbers and migration.

- **constitution**
  set of written rules made by members of a group.

- **contagious**
  disease that can be passed from one cow to another through a number of possible ways.

- **contemporaries**
  cows of the same breed that were born and raised, and

initiated their lactations during similar periods. Usually separated into two lactation groups for comparison- first lactation and all other lactations.

- **contemporary comparison**
method for estimating the transmitting ability of bulls and cows using information on contemporaries.

- **contiguous**
very near or touching.

- **continuous infusion**
the slow introduction of a fluid into a vein or artery over a period of time.

- **controlled atmosphere packaging (CAP)**
packaging method in which selected atmospheric concentrations of gases are maintained throughout storage in order to extend product shelf life. Gas may either be evacuated or introduced to achieve the desired atmosphere. Normally used for fruits and vegetables, not meat products.

- **cooperative**
a form of business owned and controlled by the people who use its services, such as a milk marketing or processing cooperative.

- **cooperative group**
a group of physicians and/or hospitals formed to treat a large number of patients in the same way so that new treatment can be evaluated quickly.

- **co-packer sourcing**
when a manufacturer or entrepreneur has a product but nowhere to manufacture it, facilities are found; these facilities are called co-packers, since they are partners in preparing the product for the marketplace.

- **core biopsy**
the removal of tissue samples with a needle to check for cancer cells.

- **corn**
any of numerous cultivated forms of a widely grown, usually tall annual cereal grass (Zea mays) bearing grains or kernels on large ears. The grains or kernels of this plant, used as food for humans and livestock or for the extraction of an edible oil or starch. Also called Indian corn, maize.

- **corned beef**

corning is a form of curing one of the several less-tender cuts of beef like the brisket, rump or round. It has nothing to do with corn. The name comes from Anglo-Saxon times before refrigeration. In those days, the meat was dry cured in coarse 'corns' of salt. Pellets of salt, some the size of kernels of corn, were rubbed into the beef to keep it from spoiling and to preserve it. Today, brining has replaced the dry salt cure, but the name 'corned beef' is still used, rather than 'brined' or 'pickled' beef. Commonly used spices that give corned beef its distinctive flavour are peppercorns and bay leaf.

- **corticosteroids**

hormones that have antitumour activity in lymphomas and lymphoid leukemias; in addition, corticosteroids (steroids) may be used for hon-none replacement and for the management of some of the complications of cancer and its treatment.

- **corynebacterium granulosum**

a bacterium that stimulates the immune system to fight cancer.

- **cottage ham**

1. a ham made from the shoulder butt end.
2. uncooked, cured, dried, smoked-or-unsmoked meat products made from a single piece of meat from the hind leg of a hog or from a single piece of meat from a pork shoulder.

- **coverall**

a brand name, but commonly referring to any of a group of hoop type barns with opaque or mylar fabric covers over a tubular steel frame.

- **cow**

1. a mature bovine female, usually having had a least one calf.
2. a mature female bovine. Usually referring to any dairy females that have borne a calf. Some may consider females having given birth only once as 'first-calf heifers' until they have a second calf.

- **cow trainer**

a tin or wire structure supported a few inches above a cow to prevent her from soiling the platform of her stall by administering a gentle electric shock if she arches her back to urinate or defecate while too far forward in the stall.

- **cow-hocked**

a condition of a cow in which the hocks are close together and the fetlocks (located just above the hoof) wide apart.

- **crampiness**

a condition among cattle, especially bulls in confinement, in which involuntary contractions of muscles of the hind legs result in pain and discomfort. Affected animals shift from foot to foot.

- **crate**

an elevated stall for a calf in an indoor facility.

- **cream**

1. milk product.
2. fat component of milk. Pura creams contain between 18 % fat for Pura Light cream and 35% fat for Pura Double thickened cream.

- **creamery**

a butter factory or, more commonly, a milk processing plant.

creatinine
a blood and urinary chemical that is used to estimate overall kidney function. It is produced by the muscles at a regular, predictable rate and excreted by the kidneys through simple filtration. If the blood creatinine level becomes too high, it often means that not enough blood is being filtered by the kidneys. A serum creatinine up to 1.5 to 2.0 is generally considered normal, but lower numbers are usually better.

creatinine clearance
a calculated value that is a good indicator of kidney function. Since the kidneys based on kidney function filter serum creatinine, it is possible to measure the total urinary creatinine passed in 24 hours and calculate how much blood the kidneys filtered in that time to produce the creatinine found in the sample. It comes out to about 100 cc's per minute. Anything below 30 cc's is considered to be kidney failure.

creep feeder
a small feeding pen built with a small entrance through which only young, small animals can enter. This prevents larger animals from crowding the smaller ones from the feed.

creutzfeldt-jacob disease (CJD)
a sporadic and rare, but fatal human disease that usually strikes people over 65. It occurs worldwide at an estimated rate of one case per million population. About 10-15% of CJD cases are inherited. A small number of cases occurred as the result of various medical treatments or procedures which inadvertently transferred the CJD agent.

crimped
rolled with corrugated rollers. The grain to which this term refers may be tempered or conditioned before crimping and may be cooled afterward.

- **crisnatol mesylate**
an investigational anticancer drug that interferes with the DNA in cancer cells.

- **critical control point**
an operation (practice, procedure, process, or location) at or by which preventive or control measures can be exercised that will eliminate, prevent, or minimise one or more hazards. Critical control points are fundamental to Hazard Analysis and Critical Control Point (HACCP) systems.

- **crohn's disease**
  1. a chronic inflammatory bowel disease of unknown origin, usually affecting the ileum, the colon, or both structures. Normal bowel segments may separate diseased segments. See also **Colitis, Ulcerative Colitis**
  2. chronic inflammation of the gastrointestinal tract, most commonly the bowel

- **crossbred**
offspring of parents of different breeds.

- **crossbred/crossbreed**
a combination of two or more breeds eg Holstein Friesian x Jersey.

- **crossbreeding**
mating animals of different breeds.

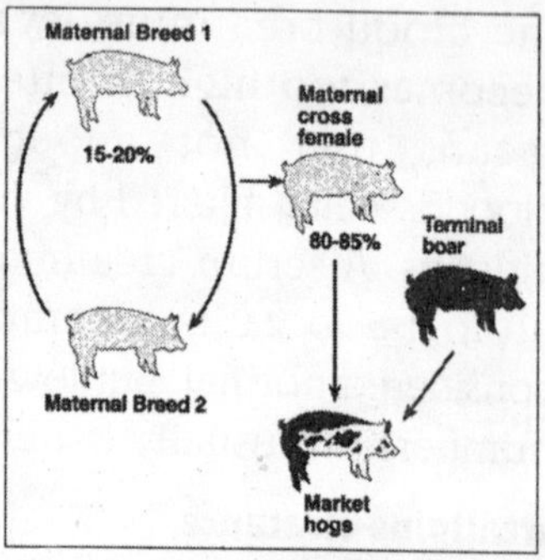

- **cross-contamination**
the transfer of harmful substances or disease-causing microorganisms to food by hands, food-contact surfaces, sponges, cloth towels and utensils that touch raw food, are not cleaned, and then touch ready-to-eat foods. Cross contamination can also occur when raw food touches or drips onto cooked or ready-to-eat foods.

- **crossing over**
exchange of genes between two paired chromosomes.

- **cross-licensing**

  legal, contractual procedure in which two or more firms with competing, similar technologies and possible conflicting patent claims strike a deal to reduce the need for legal actions to clarify who is to profit from applications of the technology.

- **crowd gate**

  a motorised or manual gate at the end of the holding pen that may be moved forward to guide cows toward the entrance to the milking parlour.

- **crude fibre (CF)**

  that portion of feedstuffs composed of cellulose, hemicellulose, lignin and other polysaccharides which serve as structural and protective parts of plants (high in forages and low in grains). Not soluble in acid or alkali detergents.

- **crude protein**

  1. crude protein, sometimes called total protein, is estimated from measuring the total nitrogen content of milk. 6.38 to express the results on a protein equivalent basis multiply nitrogen.
  2. total protein in a feed. To calculate the protein percentage, a feed is first chemically analysed for nitrogen content. Since proteins average about 16 percent (100 ö 6.25) nitrogen, the percentage of nitrogen in the analysis is multiplied by 6.25 to give the CP percentage.

- **crustacean**

  invertebrate animals with many jointed legs and a hard external shell, such as crawfish, shrimp, soft-shell crabs and prawns.

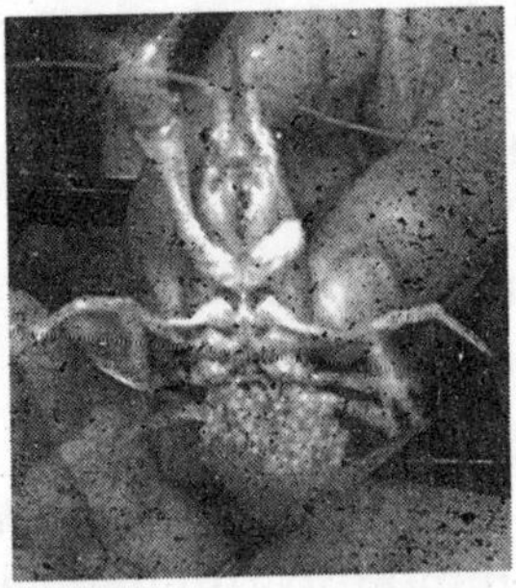

- **cryogenic**

  very low temperature freezing.

- **cryosurgery**

  freezing body tissues to sub-zero temperatures to destroy cancer cells.

- **cryotherapy**
  the use of cold to treat disease.
- **CSF**
  the abbreviation for cerebrospinal fluid, which is the liquid that surrounds the brain and spinal cord. An abbreviation for the combination chemotherapy drugs cyclophosphamide, hiotepa and carboplatin.
- **CT scan**
  short for Computerised Tomography, this refers to a computer manufactured image made from a special X-ray machine. The images are very good for soft tissues but often require contrast to look into the kidneys and the intestines. Also called a 'CAT' scan. Most kidney stones will show up well on CT scans.
- **cud**
  a bolus of regurgitated food (common to ruminants).
- **cull**
  1. to remove from the herd or flock, usually undesirable and/or inefficient (unprofitable) breeding stock that will be sent to slaughter.
  2. an animal which has past its productive life and is sent for slaughter.
  3. to remove a cow from the herd. Culling reasons in clued voluntary culling of cows for low milk production, or involuntary culling of cows for reasons of health or injury.
- **cull cow**
  a cow having been identified to be removed from the herd or having recently left.
- **culling**
  the process of eliminating nonproductive or undesirable animals. The removal of cows from a herd as a result of low production or other factors that reduce the profitability of the cow.
- **cultivation**
  preparation of land to grow crops.
- **culture**
  1. cultivation of living organisms in prepared medium, s a verb
  2. to grow in prepared medium.
- **culture**
  in microbiology, a population of microorganisms in a

growth medium or the act of growing bacteria in media for identification. A pure culture contains only organisms that initially arose from a single cell. Cultures are used in manufacturing cultured dairy products and most cheeses.

- **culture medium**
  any nutrient system for the artificial cultivation of bacteria or other cells, usually a complex mixture of organic and inorganic materials.

- **cultured dairy products**
  made by adding selected, useful bacterial cultures to milk in a controlled environment. These foods are beneficial to people with lactose intolerance and other digestive disorders. Yogurt and buttermilk are examples of cultured dairy products.

- **cultures**
  live or viable bacteria, such as acidophilus, bifidobacteria and casei cultures.

- **culvert**
  a drain or pipe that carries a watercourse under a road, railway or other built structure.

- **curd**
  the coagulated or thickened part of milk. Curd from whole milk consists of casein, fat, and whey, whereas curd from skim milk contains casein and whey but only traces of fat.

- **curd separation**
  after coagulation the milk is separated into whey and cheese curd.

- **curds**
  clotted protein formed when fresh milk is treated with rennet.

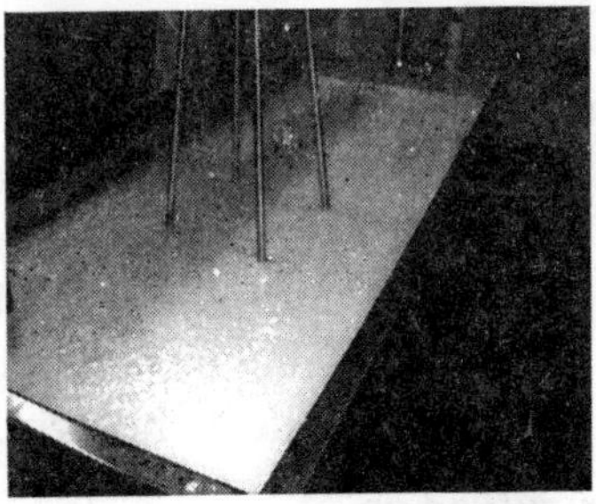

- **cure**
  a chemical agent placed in or on meat or poultry for use in preservation, flavour or colour.

- **curing**
  1. curing is the addition of salt, sodium nitrate (or saltpetre), nitrites and

sometimes sugars, seasonings, phosphates and ascorbates to pork for preservation, colour development and flavour enhancement.

2. the step in the cheese manufacturing process where cheese is allowed to sit at controlled temperatures for a fixed amount of time. The length of time cheese is allowed to cure determines its flavor and texture. Mild, medium and sharp represent cheeses with different curing or ripening times.

- **cutaneous t-cell lymphoma**

a disease in which certain cells of the lymph system (called T lymphocytes) become cancerous (malignant) and affect the skin.

- **cutting up**

any division of any carcass or part thereof, except that the trimming of carcasses or parts thereof to remove surface contaminants is not considered as cutting up.

- **CVP**

the abbreviation for the combination chemotherapy drugs cyclophosphamide, vincristine and prednisone.

- **CWT**

hundredweight or 100 pounds of milk.

- **cyclosporine**

a drug used to help prevent rejection of organ and bone marrow transplants by the body. It is also used to inhibit multidrug resistance.

- **cypripedium**

a genus of orchids that includes the lady's slipper- orchid.

- **cystectomy**

surgery to remove the bladder.

- **cystine**

an amino acid or ingredient in protein. Significant because in an inherited disorder called cystinuria, very large amounts of cystine can be excreted in the urine where it can form stones. These stones tend to be difficult to treat and require life-long therapy.

- **cystinuria**

the presence of cystine in urine. While this does not necessarily mean excess, when used in common language, it generally refers to abnormally high urinary levels of cystine.

- **cystitis**
an inflammation of the bladder which is prevalent in women. Usually caused by a urinary bladder infection.

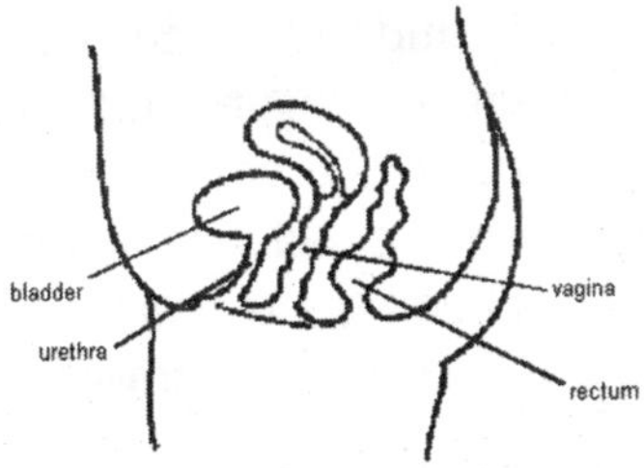

- **cystoscopy**
a procedure where a special telescope, either rigid or flexible, is passed into the penis or female urethra and enters the bladder. It is used for inspection of these organs and to place tubes, ureteral catheters and Double-J stents in the ureters or kidneys.

- **cystoscopy**
examination of the bladder with removal of tissue samples and small tumours using an instrument called a cystoscope.

- **cyto-**
referring to cell or cell plasm.

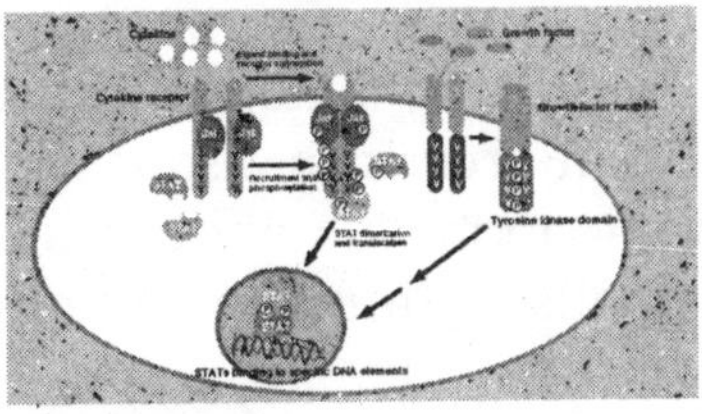

- **cytogenetics**
study of the cell and its heredity-related components, especially chromosomes.

- **cytomegalovirus**
a virus that may be carried in an inactive state by healthy individuals for life. It is a cause of severe pneumonia in bone marrow transplantation patients and patients with leukemia or lymphoma.

- **cytoplasm**
cellular material that is within the cell membrane and surrounds the nucleus.

- **cytotoxic**
able to cause cell death.

- **cytotoxic chemotherapy**
chemotherapy that kills cells, especially cancer cells.

- **cytotoxic t cells**
white blood cells that can directly destroy specific cells. T cells can be separated from other blood cells and grown in the laboratory and

then be given to the patient to destroy tumour cells. Certain drugs can also assist in the formation of cytotoxic T cells within the patient's body.

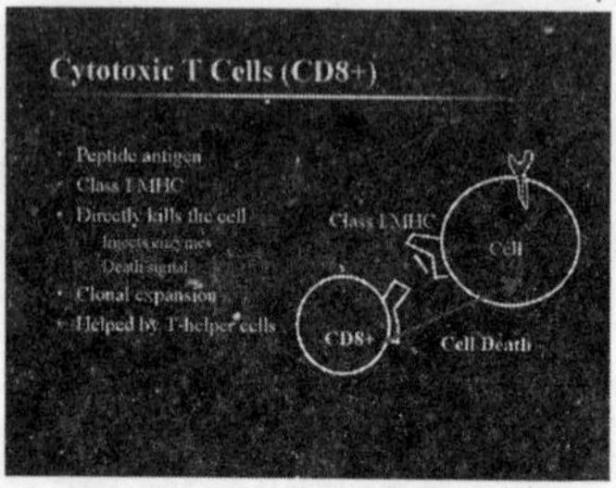

- **d-20761**

a synthetic luteinising hormone-releasing hon-none (LH-RH) antagonist that suppresses LH and sex steroid levels. Also called Cetrorelix.

- **dacarbazine**

a chemotherapy drug that combines with the DNA in cancer cells to prevent them from multiplying.

- **dactinomycin**

an investigational chemotherapy drug.

- **dairy**

1. relating to the production of milk or milk products (e.g. butter, cheese). Dairy cattle, for example, are cows kept for the production of milk as distinct from beef cattle that are raised for their meat.
2. of, for, or relating to milk and milk products typically derived from cows, however, may also come from goats, sheep, or other milk bearing animals. Dairy products include milk, cheese, ice cream, yogurt, butter, margarine, cream, and may also be included in various baked goods and recipes. See **Casein, Lactose**

- **dairy character**

physical traits that suggest high milking ability. Important indications of this include an alert, feminine head; long thin neck; openness and sharpness throughout; prominent hips and pins; sharp withers; thin thighs; flat flinty bone, and thin, pliable hide with a soft silky hair coat. See **bloom**.

■ **dairy cow**

a bovine from which milk production is intended for human consumption, or is kept for raising replacement dairy heifers.

■ **dairy goat**

any goat from which milk production is intended for human consumption, or is kept for raising replacement dairy kids.

■ **dairy steer**

a neutered male of any of the dairy cattle breeds. The 'dairy steers' are raised for meat production and usually managed like beef cattle.

■ **dam**

1. mother or female parent in a pedigree.
2. the female parent of any dairy animal.

■ **database**

livestock Improvement's database of dairy cow ancestry, production and mating statistics.

■ **daughter-herdmate comparison**

amount by which a daughter of a bull differs in yield or other traits from the average of cows of other bulls' daughters of the same breed in the same herd during the same period.

■ **days in milk (dim)**

the number of days during the present lactation that the cow has been milking, beginning with the last date of calving to the current test date.

■ **dcfi**

wt. of ingredient consumed X 100.

■ **dead livestock**

the body (cadaver) of livestock which has died otherwise than by slaughter.

■ **deciduous**

trees or shrubs that shed their leaves in the autumn. Compare with evergreen.

- **decitabine**
  a chemotherapy drug used in the treatment of cancer of the blood.
- **dedicated facility**
  an entire area or building that is used to create one type of product.
- **dedicated line**
  equipment or machinery that is used for the purpose of creating one type of product.
- **degrees celsius**
  a measure of temperature.
- **dehalogenate**
  to remove chlorine or bromine ions from common organic contaminants and thereby reduce either their toxicity or the difficulty of degrading them.
- **dehorn**
  to remove horns from animals or to treat young animals so horns will not develop.
- **dehydration**
  1. the loss of excessive water from the body. Such water loss can take place through the kidneys, the lungs or from perspiration.
  2. by ethanol, for example, leads to aggregation of the micelles.
- **deli or delicatessen style**
  this terminology has been permitted on labelling for ready-to-eat meat food products that consumers would normally expect to find in a delicatessen.
- **deliveries**
  means a subcategory of distribution, consisting of sugar delivered to end-users for consumption as sugar or for use in products containing sugar, including sugar delivered to a manufacturer under the Sugar-Containing Products Re-Export Program.
- **denitrification**
  1. the biochemical reduction of nitrate or nitrite to gaseous nitrogen, either as molecular nitrogen or as an oxide of nitrogen.
  2. the chemical or biological reduction of nitrate or nitrite to gaseous nitrogen, either as molecular nitrogen (N2) or as an oxide of nitrogen (N_2O).

■ **density**

the density of milk and milk products is used for the following

- to convert volume into mass and vice versa
- to estimate the solids content
- to calculate other physical properties (e.g. kinematic viscosity)

Density, the mass of a certain quantity of material divided by its volume, is dependant on the following

- temperature at the time of measurement
- temperature history of the material
- composition of the material (especially the fat content)
- inclusion of air (a complication with more viscous products)

With all of this in mind, the density of milk varies within the range of 1027 to 1033 kg m(-3) at 20° C.

■ **deoxyribonucleic acid (DNA)**

1. the molecule that carries the genetic information for most living systems. The DNA molecule consists of four bases (adenine, cytosine, guanine and thymine) and a sugar-phosphate backbone, arranged in two connected strands to form a double helix. See **Complementary DNA**, **Double helix**.

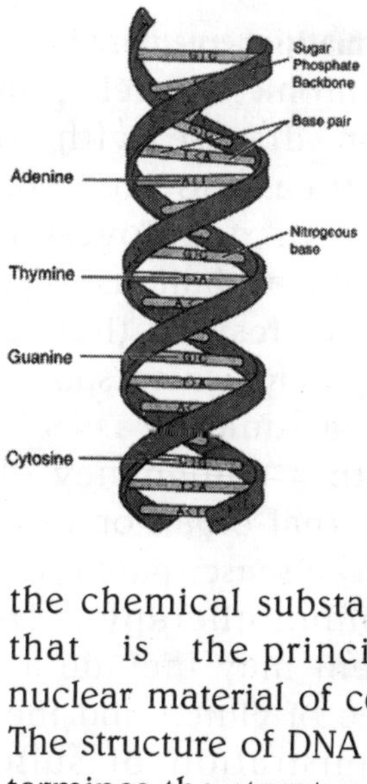

2. the chemical substance that is the principal nuclear material of cells. The structure of DNA determines the structure of ribonucleic acid, which, in turn, determines the structure of proteins of the cell.

■ **depofoam-encapsulated cytarabine**

the anticancer drug cytarabine inside small particles of a synthetic material called DepoFoam. This method of administering cytarabine results in its slow release over time.

- **depsipeptide**
  an experimental anticancer drug.
- **dermatitis**
  inflamation of the skin.
- **dermatitis herpetiformis**
  a chronic, severely pruritic skin disease with symmetrically located groups of red, papulovesicular, vesicular, bullous, or urticarial lesions that leave hyperpigmented spots. It is occasionally associated with a malignancy of an internal organ or with celiac disease, patch, or IgA immunotherapy. Treatment may include a diet free of gluten and the administration of sulfone, dapsone, sulfapyridine, or antipruritic drugs. See **gluten-free diet.**
- **desferrioxamine**
  a drug that inhibits tumour cell growth by preventing the nutrient iron from being metabolised.
- **dexamethasone**
  a synthetic hormone that may be used to relieve some of the side-effects of cancer.
- **dexrazoxane**
  a drug used to protect the heart from drugs used in cancer treatment. Also called Zinecard.
- **dextral**
  right handed. (The way in which the shells coil in some snail species, for example, may be termed 'dextral'.)
- **diabetes**
  a clinical condition characterised by the excessive excretion of urine. The excess may be caused by a deficiency of antidiuretic hormone (ADH), as in diabetes insipidus, or it may be the polyuria resulting from the hyperglycemia occurring in diabetes mellitus.
- **diafiltration**
  general term for 'through filtration', filter is not refined in any particular way.
- **diagnostic**
  a product used for the diagnosis of disease or medical condition. Both monoclonal antibodies and DNA probes are useful diagnostic products.

- **diagnostic procedure**
  a method used to identify a disease.

- **dialysis**
  the process in which a machine filters and cleans the blood. An artificial kidney machine.

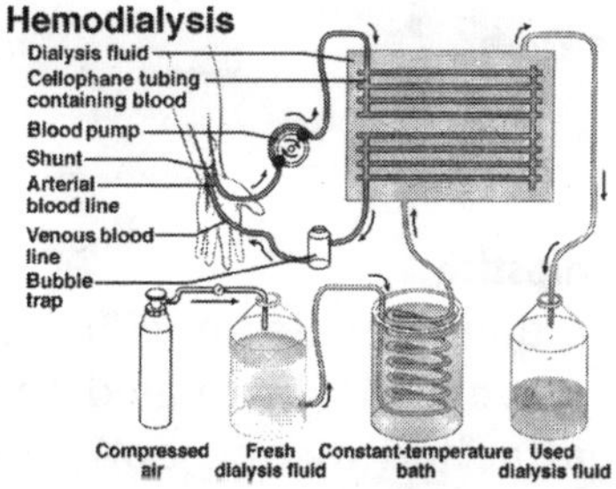

- **diaziquone**
  an anticancer drug that is able to cross the blood-brain barrier and kill cancer cells in the central nervous system. Also called AZ-Q.

- **dicoumarol**
  a chemical compound found in spoiled sweet clover and lespedeza hays. It is an anticoagulant and can cause internal haemorrhages when ingested by cattle.

- **differentiation**
  the process of biochemical and structural changes by which cells become specialised in form and function.

- **difluoromethylornithine**
  an investigational drug that has been shown to prevent cancer in animals. Also called DFMO.

- **digestibility**
  that percentage of food ingested which is absorbed into the body as opposed to that which is excreted as feces.

- **digestible energy (DE)**
  that portion of energy of a feed that can be digested or absorbed into the body by an animal.

- **digestible protein (DP)**
  the amount of protein of feed that is absorbed by the digestive tract; it may be computed using the formula: Percent DP = percent crude protein of feed x digestion coefficient for protein in the feed.

- **digestion coefficient**
  the digestion coefficient of feed ingredients (DCFI) may be calculated using the formula: Wt. of ingredient consumed — Wt. of undigested ingredient in feces.

- **diglycerides, monoglycerides**
  emulsifying agents for rendered fats.

- **diluent**
  a solution used to dilute the concentration of a liquid.

- **dioxins**
  a group of chemical compounds that share certain similar chemical structures and biological characteristics. Dioxins are present in the environment all over the world. Within animals, dioxins tend to accumulate in fat. About 95% of the average person's exposure to dioxins occurs through consumption of food, especially food containing animal fat. Scientists and health experts are concerned about dioxins because studies have shown that exposure may cause a number of adverse health effects.

- **dip**
  degradable intake protein. That protein or nitrogen degraded in the rumen by micoorganisms into microbial protein or freed as ammonia.

- **diploid**
  a cell with two complete sets of chromosomes. Compare Haploid.

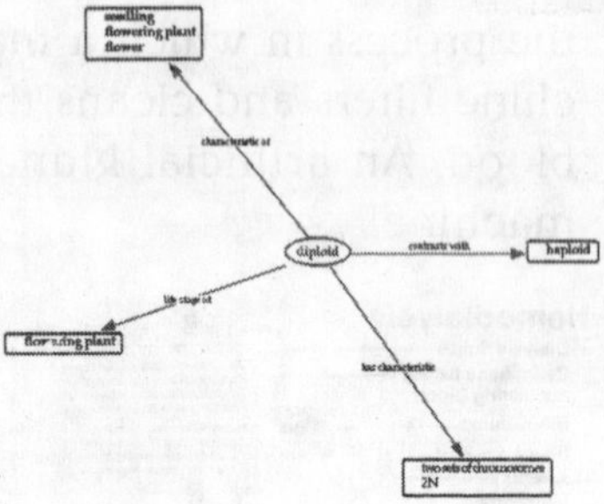

- **dipstick**
  a thin plastic strip with various chemically treated pads attached at one end. The strip with its attached pads is dipped in the urine to be tested. The various pads turn colours depending on various chemical or cellular characteristics of the urine being tested. Acid content and concentration are two of the more common tests commonly performed with urinary dipsticks. Disease specific dipsticks for kidney stone patients with identified chemical problems are under development.

- **diptera**
  a large and varied order of insects that is made up of all two- winged flies.

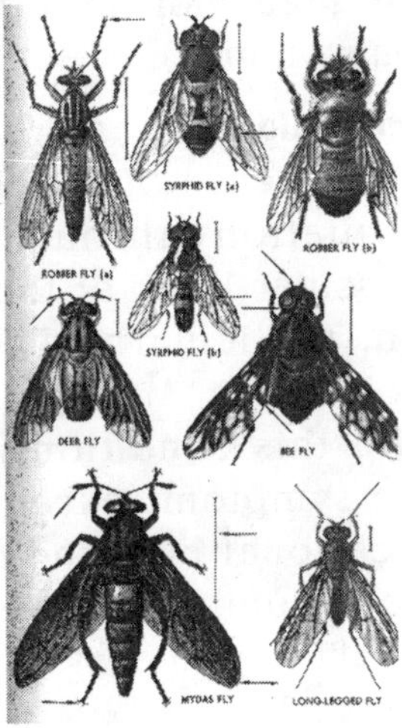

- **dipterist**
  a person that studies two-winged flies.

- **dipyridamole**
  a drug that enhances the ability of methotrexate to kill tumour cells.

- **direct microscopic somatic cell count (DMSCC)**
  microscopic count of the actual number of somatic cells in milk.

- **direct-consumption sugar**
  means any sugar which is not to be further refined or improved in quality, whether such sugar is principally of crystalline structure or is liquid sugar, edible molasses, sugar syrup or cane syrup.

- **disease progression**
  when cancer continues to grow or spread.

- **disinfectant**
  a substance that destroys harmful germs.

- **dispersal**
  the tendency of an organism to move away from the place where it was born or the site where it breeds.

- **disposition**
  a food manufacturer's action to correct a situation leading to a recall such as relabelling, reworking, or destroying product.

- **distant cancer**
  cancer that has spread to distant organs or distant lymph nodes.

- **distribution**
  means the sale or other disposition of sugar, including (but not limited to) the forfeiture of sugar to the CCC and the disposition of sugar for retail sale, further processing or refining, production of alcohol or feed or exportation.

- **disulfiram**
  a drug that slows the metabolism of retinoids, allow-

ing them to act over a longer period of time.

- **disulphide bonds**

alpha(s1) and ß-caseins do not have any cysteine residues. If any S-S bonds occur within the micelle, they are not the driving force for stabilisation.

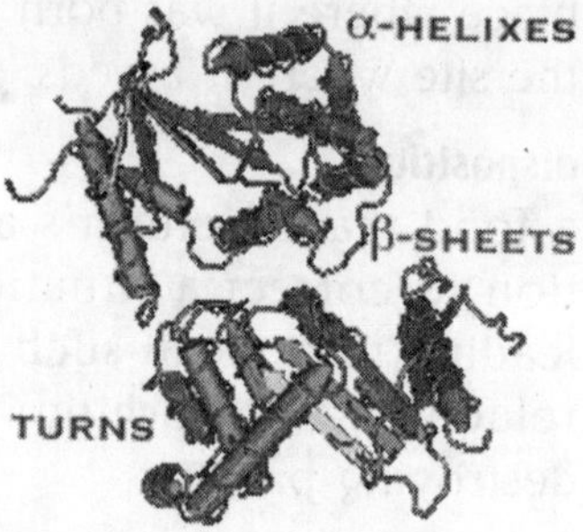

- **diuretic**

1. proper name for a 'water pill.' A medication that causes the kidney to make more urine. Often used to reduce swelling and bloating. Some diuretics have unique effects on calcium.
2. a drug that increases the production of urine.

- **diver**

member of a family of birds (Gaviidae) that dive from the surface of water to feed on fish, invertebrates or other prey. Includes birds such as the 'great northern diver' (Gavia immer).

- **diverticulosis**

the presence of pouchlike herniations through the muscular layer of the colon, particularly the sigmoid colon. Most patients with this condition have few symptoms except for occasional bleeding from the rectum. An increase in the dietary fiber is recommended.

- **DNA**

1. stands for Deoxyribonucleic Acid. This is the genetic code material found in virtually every living cell. When this DNA is digested, it produces urine, which ultimately forms uric acid.

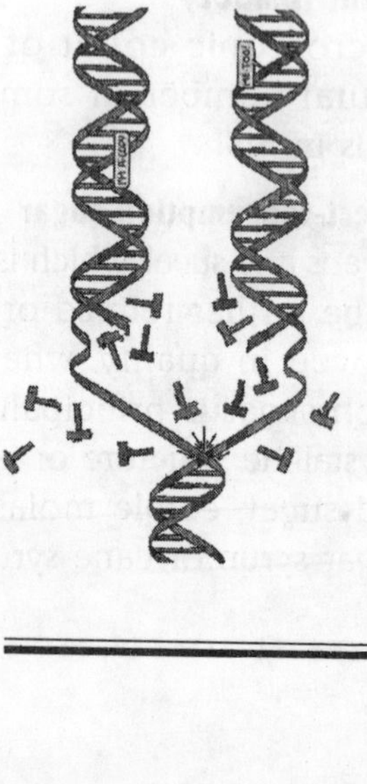

2. a molecule that carries genetic information.
3. deoxyribonucleic acid. The basic unit of DNA is the nucleotide (containing one of four bases). A DNA strand can be made up of thousands of nucleotides joined by a backbone of phosphate sugars. Chromosomes typically are constructed of two strands of DNA twisted together in the shape of a double helix. Specific nucleotide sequences define genes, which encode instructions for cell function and products (including proteins). The human genome contains 100,000 genes, or 3 x 109 base pairs; the simplest bacterium contains about 2,000 genes, or 2 million base pairs.

▪ **dna 'fingerprint'**
the chemical/physical segments of DNA that are 100 to 1,000 base pairs long. DNA fingerprints illuminate the genetic differences between and among individuals and are used in fields ranging from agricultural research to law enforcement. The extent of the variation depends on the target region of DNA being compared. Some regions appear to be common to all forms of life, while others are highly variable.

▪ **DNA chip**
a small piece of glass or silicon that has small pieces of DNA arrayed on its surface.

▪ **DNA fingerprinting**
the use of restriction enzymes to measure the genetic variation of individuals. This technology is often used as a forensic tool to detect differences or similarities in blood and tissue samples at crime scenes.

▪ **DNA hybridisation**
the formation of a double-stranded nucleic acid molecule from two separate strands. The term also applies to a molecular technique that uses one nucleic acid strand to locate another.

▪ **DNA library**
a collection of cloned DNA fragments that collectively represent the genome of an organism.

- **DNA polymerase**
  an enzyme that replicates DNA. DNA polymerase is the basis of PCR — the polymerase chain reaction.

- **DNA probe**
  1. a small piece of nucleic acid that has been labelled with a radioactive isotope, dye or enzyme and is used to locate a particular nucleotide sequence or gene on a DNA molecule.
  2. a fragment of DNA used to recognise a specific complementary DNA sequence, or gene(s). Probes can be employed, for example, to bind to the genetic material of microbes for purposes of detection, identification, or, in some cases, inactivation.

- **DNA repair enzymes**
  proteins that recognise and repair certain abnormalities in DNA.

- **DNA sequence**
  the order of nucleotide bases in the DNA molecule.

- **DNA sequencing**
  a technique for determining the order of nucleotide bases in a segment of DNA. This technique was made possible by the Nobel Prize-winning development of molecular cloning techniques for replicating genetic material. Automated sequencing technology is available now, and researchers are beginning to sequence entire genomes of free-living organisms, such as bacteria, yeast (single-celled fungi), and humans.

- **DNA vaccines**
  pieces of foreign DNA that are injected into an organism to trigger an immune response.

- **docetaxel**
  one of a type of chemotherapy agents called taxanes that block microtubule formation during cell division. Also called Taxotere.

- **dock**
  to remove a cow's tail. This practice may keep cows udders cleaner, but may also result in cows being less content, especially in fly season.

- **docking**
  removing the tail for sanitation purposes.

■ **doe**

1. a female goat, deer or rabbit.
2. a sexually mature female goat.

■ **doeling**

a female goat from one to two years old, or from one year old until first parturition.

■ **dominant**

describes a gene that covers up the physical expression of its paired allele or recessive gene. For example, black-and-white, Holsteins (B) is dominant over red-and-white (b). Thus, BB and Bb Holsteins are black-and-white, whereas bb individuals are red-and-white.

■ **donor dams**

genetically superior cows from which embryos are collected and transferred to recipients to allow these cows to produce a greater number of offspring than possible in a normal reproduction scheme.

■ **donor site**

the site from which individuals of a species are taken in order to be introduced to another place (the 'receptor site').

■ **dose-rate**

the strength of a treatment given over a period of time.

■ **double helix**

a term often used to describe the configuration of the DNA molecule. The helix consists of two spiraling strands of nucleotides (a sugar, phosphate and base) joined crosswise by specific paring of the bases. See Deoxyribonucleic acid, Base, Base pair.

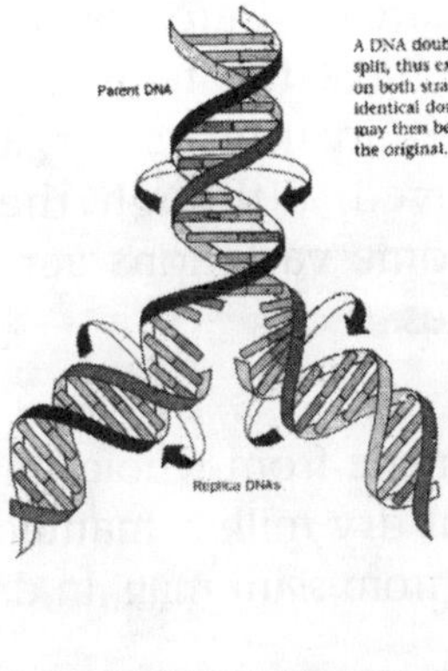

A DNA double helix may be split, thus exposing bases on both strands. Two new identical double helices may then be formed from the original.

- **double standardisation**
this refers to standardisation of milk for both fat and non-fat solids (or protein).

- **double-blinded**
a doubled-blinded trial is a clinical trial in which neither the medical staff nor the patient knows if the patient is receiving the investigational drug or the placebo.

- **downer (or downed animals)**
commonly used term for an animal that is unable to rise and walk.

- **downer cow**
a cow unable to arise due to disease or injury.

- **dredging**
removal of mud or silt from a water body or water-course.

- **dressed weight**
the weight of a chilled animal carcass. Carcasses typically have the feet, head, hide, and internal organs removed, although there are some variations across species.

- **dried milks**
it is made from whole milk, nonfat dry milk is manufactured from skim milk. In drying milks, about 95 to 98 percent of the water is removed by concentrating the solids in vacuum pans and then drying by spraying milk into a chamber of hot air or onto a slowly revolving heated drum under vacuum.

- **drug delivery**
the process by which a formulated drug is administered to the patient. Traditional routes have been oral or intravenous perfusion. New methods deliver through the skin with a transdermal patch or across the nasal membrane with an aerosol spray.

- **dry**
describes a non-lactating female. The dry period of cows is the time between lactations (when a female is not secreting milk).

- **dry aged**
fresh Meat is held (without vacuum packing) for various periods of time (usually 10 days to 6 weeks) under controlled temperatures (34°F to 38°F), humidity, and airflow to avoid spoilage and ensure flavour enhancement, tenderness, and palatability.

■ **dry cow**

1. a cow that is not lactating or secreting milk after it has completed a lactation period following calving.

2. any cow that has calved once and is not producing milk.

■ **dry curing**

dry curing is the process used to make country hams and prosciutto. Fresh meat is rubbed with a dry cure mixture of salt and other ingredients. Dry curing produces a salty product.

■ **dry date**

first calendar day the cow is not milked. See **dry period**.

■ **dry matter (dm)**

the moisture-free content of feeds.

■ **dry off**

to change a lactating animal to a non-lactating one. Usually done 6 to 8 weeks prior to freshening.

■ **dry period**

non-lactating days between lactations. This non-lactating time is generally a 5- to 6-week rest period before freshening.

■ **drying off**

the process of using certain management practices, such as reducing milking frequency and changing feed rations, to stop milk production. A dry cow is not lactating.

■ **drylot**

a relatively small area in which cattle are confined indefinitely as opposed to being allowed to have free access to pasture.

■ **drystone wall**

a stone wall built without mortar.

■ **dying, diseased, or disabled livestock**

livestock which has or displays symptoms of having any of the following

1. Central nervous system disorder
2. Abnormal temperature (high or low)
3. Difficult breathing
4. Abnormal swellings
5. Lack of muscular coordination
6. Inability to walk normally or stand
7. Any of the conditions for which livestock is required to be condemned

on ante-mortem inspection in accordance with the regulations.

- **dysphasia**
difficulty swallowing, commonly associated with obstructive or motor disorders of the oesophagus.

- **dysplasia**
abnormal changes in the way tissue cells look under a microscope.

- **dyspnea**
difficult or laboured respiration.

- **dystocia**
abnormal or difficult labor (parturition), causing difficulty in delivering the fotus and placenta.

- **e. coli o157:h7 (escherichia coli o157:h7)**
a bacterium that lives harmlessly in the intestines of animals such as cattle, reptiles, and birds. However, in humans the bacterium, which can be transmitted by foods, animal contact and drinking water, can cause bloody diarrhoea, and also lead to haemolytic uremic syndrome (HUS), a life threatening disease. Although other generic strains of E. coli are thought to be harmless to humans, the O157:H7 strain is particularly virulent and dangerous.

- **eartag**
a tag that generally is attached to the ear of a dairy cow for easy and accurate identification.

- **earthen storage basin**
an earthen pond or basin for storing animal waste. Differs from an anaerobic treatment lagoon in that wastes are not stored for and extended period and microbial breakdown or treatment does not occur.

- **echolocation**
the determination of the position of an object from the direction of an echo reflected from the object and/or the time taken for the echo to return. Bats to locate insect prey use this technique.

- **ecology**
the study of living organisms in relation to their environment.

- **ecosystem**
a unit consisting of living and non-living components

interacting to produce a stable system (e.g. a woodland or grassland).

- **edema**
  1. swelling caused by excess fluid in body tissues.
  2. the presence of abnormally large amounts of fluid in the intercellular tissue spaces of the body, as in a swelling of mammary glands commonly accompanying the initiation of the birthing process in many farm animals.
  3. the presence of abnormally large amounts of fluid in the intercellular tissue spaces of the body, as in swelling of mammary glands commonly accompanying parturition in many farm animals.

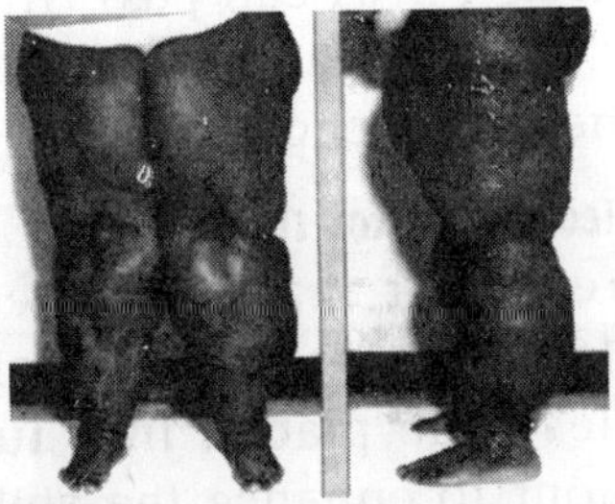

- **edible**
intended for use as human food.

- **edible molasses**
means molasses which is not to be further refined or improved in quality and which is to be distributed for human consumption, either directly or in molasses-containing products.

- **ef5**
a drug that is used to plan cancer treatment by measuring oxygen levels in tumour cells.

- **efficient cows**
cows which convert grass to milk and which conceive and calve each year.

- **effluent**
the liquid outflow or discharge of a waste treatment process.

- **eflornithine**
an investigational drug that reduces the growth of tumour cells and may stop abnormal cells from turning into cancer cells.

- **egg products**
eggs that are removed from their shells for processing. The processing of egg products includes breaking eggs, filtering, mixing, stabilising, blending, pasteurising, cool-

ing, freezing or drying, and packaging. Egg products include whole eggs, whites, yolks and various blends with or without non-egg ingredients that are processed and pasteurised and may be available in liquid, frozen and dried forms.

- **ejaculate**
  a collection of sperm.
- **elections**
  group members choosing their leaders.
- **electrodialysis**
  separation of components of a substance speeded by the application of an electric current over a membrane.

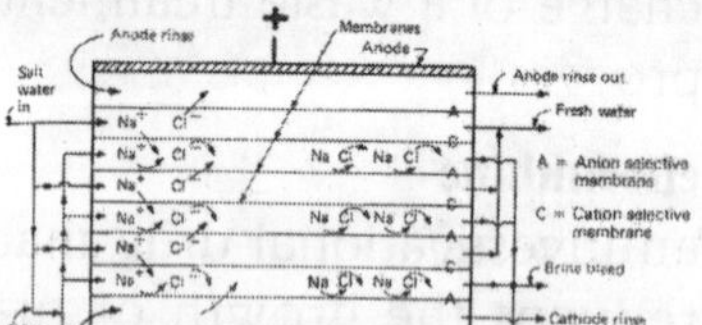

- **electrohydraulic lithotripsy (EHL)**
  a form of kidney stone fragmentation utilising a small electric probe. A small spark is created at the tip of the probe, which is placed just next to the stone to be treated. The spark generates a small shock wave, which can fragment most stones. Usually used through some type of telescope.
- **electronic feeders**
  stations in which cows are fed specified amounts of feed by a computer that recognises their unique electronic identification transponders.
- **electrophoresis**
  a technique for separating different types of molecules based on their patterns of movement in an electrical field.
- **electroporation**
  the creation of reversible small holes in a cell wall or membrane through which foreign DNA can pass. This DNA can then integrate into the cell's genome.
- **electroporation therapy**
  treatment that generated electrical pulses through an electrode placed in a tumour to enhance the ability of chemotherapy drugs to enter tumour cells. Also called EPT.

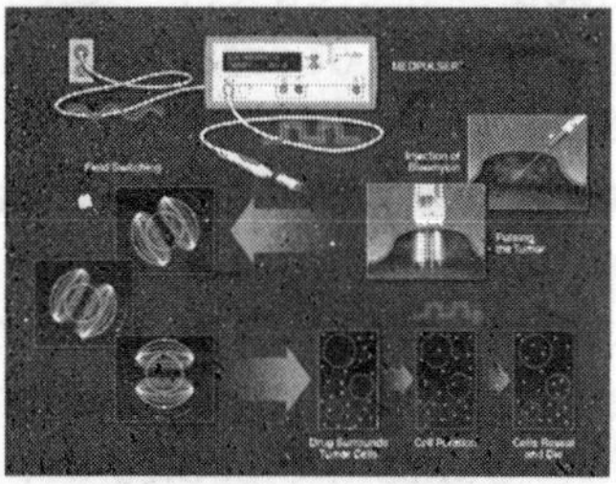

- **electrostatic interactions**
  some of the subunit interactions may be the result of ionic bonding, but the overall micellar structure is very loose and open.

- **elimination diet**
  a temporary diet used by individuals seeking to determine food allergies and intolerances where all suspected foods are eliminated for a period of time and then gradually reintroduced to note response or reactions.

- **elite sire**
  the best of the best.

- **ELUD**
  see **extra label use drug**.

- **emaci ation**
  a wasted condition of the body, great losses of body weight or condition.

- **embryo recipient**
  any cow or heifer that serves as a surrogate mother and carries the embryo of another cow throughout the development of the embryo and birth of the calf.

- **embryo transfer**
  modern technology whereby multiple fertilised eggs (ova) are flushed from the donor's uterus, and are transferred to a recipient animal that serves as a surrogate mother. The fertilised ova may be frozen and stored indefinitely before they are thawed and transferred to recipients.

- **embryonic stem cells**
  cells that can give rise to any type of differentiated cell. They can be derived from two sources: the inner cell mass from a blastocyt or the primordial germ cells (eggs and sperm) of an older embryo.

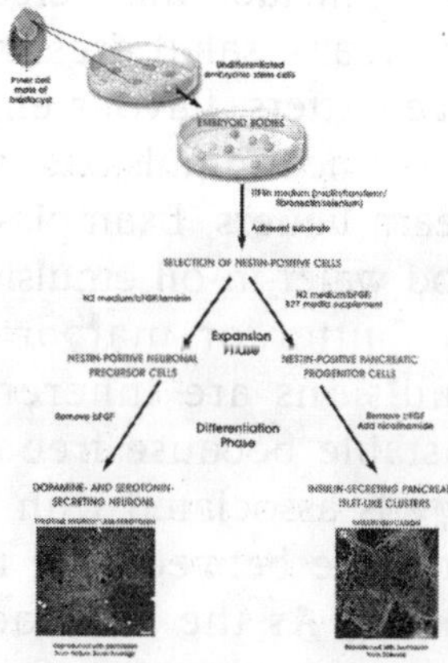

- **emulsification**
  the process of dispersing one liquid in other immiscible liquid(s).

- **emulsifier**
  a substance added to products, such as meat spreads, to prevent separation of product components to ensure consistency. Examples of these types of additives include lecithin and mono- and diglycerides.

- **emulsion**
  liquid droplets dispersed in another immiscible liquid. The dispersed phase droplet size ranges from 0.1 - 10 μ m. Important oil-in-water food emulsions, ones in which oil or fat is the dispersed phase and water is the continuous phase, include milk, cream, ice cream, salad dressings, cake batters, flavour emulsions, meat emulsions, and cream liquers. Examples of food water-in-oil emulsions are butter or margarine. Emulsions are inherently unstable because free energy is associated with the interface between the two phases. As the interfacial area increases, either through a decrease in particle size or the addition of more dispersed phase material, i.e. higher fat, more energy is needed to keep the emulsion from coalescing. Some molecules act as surface-active agents (called surfactants or emulsifiers) and can reduce this energy needed to keep these phases apart.

- **enabling environment**
  external factors for a group.

- **enclosure**
  an area of land that is walled or fenced.

- **endemic**
  only occurring in a particular area, country or region. For example, the Lancaster whitebeam is endemic to the Morecambe Bay area.

- **endocrine cancer**
  cancer that occurs in endocrine tissue, the tissue in the body that secretes hormones.

- **endometrial disorder**
  abnormal cell growth in the endometrium (the lining of the uterus).

- **endoparasite**
  a parasite that lives inside its host rather than externally.
- **endoscopy**
  the use of an endoscope, a flexible, lighted tube, for examining the inside of the body.

- **endostatin**
  an endogenous protein that blocks the proliferation of blood vessels.
- **ensilage**
  a green chop (forage) preserved by fermentation in a silo, pit, or stack, usually in chopped form. Also called silage.
- **enteric hyperoxaluria**
  condition usually associated with some form of chronic diarrhea and fat malabsorption. Very high levels of urinary oxalate are produced by a lack of intestinal calcium, together with chronic diarrhea and very low urinary calcium.
- **entomologist**
  a person who studies insects.
- **entomology**
  the study of insects.
- **environment**
  the natural world of land, sea, air, plants and animals. The term is also used in a wider sense to mean all the surroundings in which people live and work.
- **environmental**
  derived from the animal's environment, bedding, housing, etc.
- **enzyme**
  a protein catalyst that facilitates specific chemical or metabolic reactions necessary for cell growth and reproduction.
- **enzyme coagulation**
  chymosin, or rennet, is most often used for enzyme coagulation. During the primary stage, rennet cleaves the Phe(105)-Met(106) linkage of kappa-casein resulting in the formation of the soluble CMP which diffuses away from the micelle and

para-kappa-casein, a distinctly hydrophobic peptide that remains on the micelle. The patch or reactive site, as illustrated in the above image, that is left on the micelles after enzymatic cleavage is necessary before aggregation of the paracasein micelles can begin.
During the secondary stage, the micelles aggregate. This is due to the loss of steric repulsion of the kappa-casein as well as the loss of electrostatic repulsion due to the decrease in pH. As the pH approaches its isoelectric point (pH 4.6), the caseins aggregate. The casein micelles also have a strong tendency to aggregate because of hydrophobic interactions. Calcium assists coagulation by creating isoelctric conditions and by acting as a bridge between micelles. The temperature at the time of coagulation is very important to both the primary and secondary stages. With an increase in temperature up to 40° C, the rate of the rennet reaction increases. During the secondary stage, increased temperatures increase the hydrophobic reaction. The tertiary stage of coagulation involves the rearrangement of micelles after a gel has formed. There is a loss of paracasein identity as the milk curd firms and syneresis begins.

- **enzyme-linked immunosorbent assay (ELISA)**

1. a technique for detecting specific proteins by using antibodies linked to enzymes.
2. a sensitive, inexpensive assay technique involving the use of antibodies coupled with indicators (e.g., enzymes linked to dyes) to detect the presence of specific substances, such as enzymes, viruses or bacteria.

- **enzymes**

enzymes are a group of proteins that have the ability to catalyse chemical reactions and the speed of such reactions. The action of enzymes is very specific. Milk contains both indigenous and exogenous enzymes. Exogenous enzymes mainly consist of heat-stable enzymes produced by psychrotrophic bacteria: lipases, and proteinases. There are many indigenous enzymes that have been isolated from milk. The most significant group are the hydrolases

- lipoprotein lipase
- plasmin
- alkaline phosphatase.

▪ **enzymes**
protein catalysts, produced by living cells, that facilitate specific chemical or metabolic reactions.

▪ **ependymal tumours**
brain tumours that begin in the ependyma, the cells that line the passageways in the brain where special fluid that protects the brain and spinal cord (called cerebrospinal fluid) is made and stored. An ependymoma is an ependymal tumour. The different kinds of ependymal tumours are identified by the way the cells look under a microscope

▪ **epidemic**
the rapid spreading of a disease so that many animals or people have it concurrently.

▪ **epidemiology**
study of the distribution of disease, or other health-related conditions and events in human or animal populations, in order to identify health problems and possible causes.

▪ **epidural**
an injection given into the lower back.

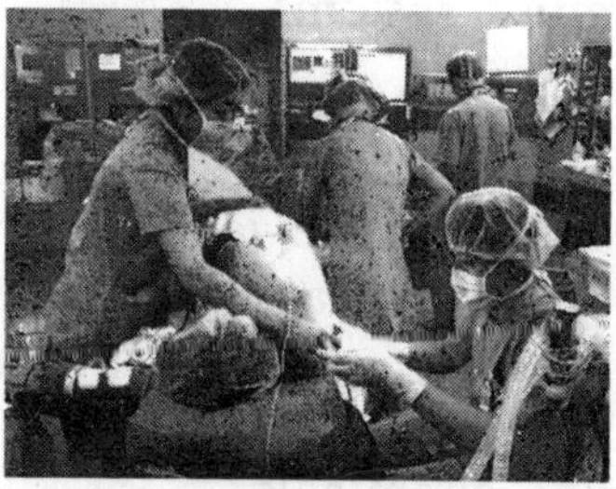

▪ **epinephrine**
a hormone. Also called adrenaline.

▪ **epithelial**
the cells that line the internal and external surfaces of the body.

▪ **epizootic**
designating a widely diffused disease of animals spreading rapidly and affecting many individuals of a kind concurrently in any region, thus corresponding to an epidemic in man.

- **epoetin alfa**
  a drug that stimulates the productioh of red blood cells.

- **epstein-barr virus**
  a common virus that remains dormant in most people. It has the potential to cause B cells to multiply uncontrollably in some patients with weakened immune systems.

- **equipment sanitisation**
  the removal of microorganisms and fat, protein, and mineral residues in milking equipment through use of water, heat and chemicals.

- **equipment sourcing**
  helping clients find the equipment they need for their plants.

- **equivalence**
  a term applied by the Uruguay Round Agreement on the Application of Sanitary and Phytosanitary (SPS) Measures. WTO Member countries shall accord acceptance to the SPS measures of other countries (even if those measures differ from their own or from those used by other Member countries trading in the same product) if the exporting country demonstrates to the importing country that its measures achieve the importer's appropriate level of sanitary and phytosanitary protection.

- **ergosterol**
  a plant sterol which, when activated by ultraviolet rays, becomes vitamin D2 , also called provitamin D2 and ergosterin.

- **eructation**
  the act of belching or casting up gas from the stomach.

- **erythropoietin (EPO)**
  1. a hormone that stimulates peripheral stem cells in the bone marrow to produce red blood cells.
  2. a protein that boosts production of red blood cells. It is clinically useful in treating certain types of anaemia.

- **escarpment**
  the steep slope that terminates a plateau or any level upland area of land.

- **escherichia coli (e. coli)**
  a bacterium that inhabits the intestinal tract of most vertebrates. Much of the work using recombinant DNA techniques has been carried out with this organism because it has been genetically well characterised.

- **escutcheon**
  the part of a cow that extends upward just above and back of t he udder where the hair turns upward in contrast to the normal downward direction of hair. Also called milk mirror.

- **estramustine**
  an anticancer drug that may stop the growth of cancer cells and eventually destroy them.

- **estrus (oestrus, estrous)**
  the recurrent, restricted sexual receptivity (heat) in female mammals, commonly marked by intense sexual urge. Estrous pertains to the entire cycle of reproductive changes in the non-pregnant female animal.

- **etanidazole**
  a drug that increases the effectiveness of radiation therapy.

- **ether extract (EE)**
  fatty substances or lipids of foods and feeds that are soluble in ether. European breeds Those native to Great Britain or continental Europe such as Charolais, Simmental, Limousin, Jersey, Holstein and Guernsey.

- **ethynyluracil**
  a chemotherapy drug that increases the effectiveness of fluorouracil.

- **etidronate**
  belongs to the group of drugs known as bisphosphonates that are used as treatment for hypercalcemia (abnormally high levels of calcium in the blood) and for cancer that has spread to the bone (bone metastases).

- **etoposide**
  belongs to a group of drugs known as plant alkaloids. Also known as VP-16.

- **EU habitats and species directive**
  this is the formal title of the habitats directive.

- **EU habitats directive**
  see **habitats directive**

- **eukaryote**

  a cell or organism containing a true nucleus, with a well-defined membrane surrounding the nucleus. All organisms except bacteria, viruses and cyanobacteria are eukaryotic. Compare Prokaryote.

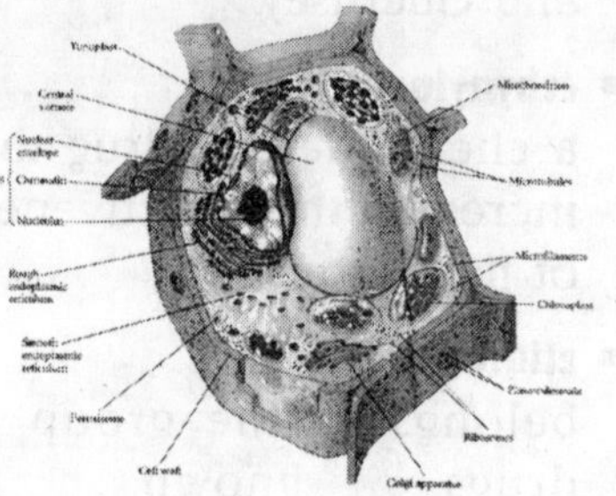

- **eutrcphic**

  nutrient enriched.

- **eutrophication**

  the process whereby an ecosystem becomes nutrient enriched. Usually the term is used in relation to freshwater aquatic systems. It may be caused by pollution from agricultural or urban run-off.

- **evaluable disease**

  extent of disease that cannot be measured directly by size of the tumour but can be evaluated by other methods specific to a particular clinical trial

- **evaporated milk**

  1. a canned whole milk concentrate formed by evaporating enough water from fresh whole milk to reduce the volume in half. The concentrate is then homogenised, fortified with vitamin D, packed in cans and sterilised by heat. Evaporated milk contains at least 25.5% milk solids and 7.5% milkfat. It may be used straight from the can in sauces and gravy and in many other recipes.
  2. the liquid food obtained by partial removal of water from milk. The MF and TMS contents are not less than 7.5 and 25.5 percent, respectively. It contains 25 IU vitamin D per ounce and is homogenised. The food is sealed in a container and is so processed by heat, either before or after sealing, as to prevent spoilage.

- **evergreen**

  trees or shrubs that bear leaves all year round. Compare with deciduous.

- **ewe**
  a female sheep.

- **exocrine (eccrine)**
  secreting outwardly, into or through a duct.
- **exon**
  in eukaryotic cells, that part of the gene that is transcribed into messenger RNA and encodes a protein. See **Intron**.
- **exotic newcastle disease**
  a contagious and fatal viral disease affecting all species of birds. Exotic Newcastle disease is one of the most infectious diseases of poultry in the world. A death rate of almost 100 percent can occur in unvaccinated poultry flocks.
- **experimental animal**
  any animal used in any research investigation involving the feeding or other administration of, or subjection to, an experimental biological product, drug, or chemical or any non-experimental biological product, drug, or chemical used in a manner for which it was not intended.
- **expression**
  in genetics, manifestation of a characteristic that is specified by a gene. With hereditary disease, for example, a person can carry the gene for the disease but not actually have the disease. In this case, the gene is present but not expressed. In industrial biotechnology, the term is often used to mean the production of a protein by a gene that has been inserted into a new host organism.
- **extant**
  existing or surviving.
- **extra label use drug**
  an antibiotic or other chemical used on the advice of a veterinarian in a dosage, route of administration, for a disease, or in some other manner not included on the approved printed package label.
- **extracorporeal shock wave lithotripsy (ESWL)**
  this refers to lithotripsy, and is sometimes called

ESWL. This is a large machine that generates a powerful shock wave using one of several different technologies. The generator is outside the body but the shock waves are focused at a point where the stones are located. Probably the most commonly used tool to fragment kidney stones today.

- **extra-label drug use**
an antibiotic or other chemical used on the advice of a veterinarian in a dosage, route of administration, for a different disease or in some other manner not included on the approved printed package label.

- **extremophiles**
microorganisms that live at extreme levels of pH, temperature, pressure and salinity.

- **extroversion**
the process of exposure of tissues inside the streak canal of cows' teats. It is usually caused by prolonged milking and excessive milking vacuum or inadequate pulsation.

- **FAC**
the abbreviation for the combination chemotherapy drugs fluorouracil, doxorubicin, and cyclophosphamide. This combination is also known as CAF.

- **factor viii**
a large, complex protein that aids in blood clotting and is used to treat haemophilia.

- **false heat**
the display of estrus by a female animal when she is pregnant.

- **family**
a unit used in the classification of animals and plants. It consists of a group of closely-related genera or, more rarely, a single genus.

- **fanconi anemia**
a rare and often fatal inherited disease in which the bone marrow fails to produce red blood cells, white blood cells, platelets, or a combination of these cells. The disease may transform into myelodysplastic syndrome or leukemia.

- **FAO**
  Food and Agriculture Organisation of the United Nations.
- **farm-to-table continuum**
  a multi-step journey that food travels before it is consumed.
- **farrow**
  to give birth to a litter of pigs.
- **fat malabsorption**
  condition in which fat is not digested normally leading to chronic diarrhoea and excessive oxalate absorption.
- **fat-corrected milk**
  the estimated quantity of milk calculated on an equivalent energy basis. It is a means of evaluating milk production records of different dairy animals and breeds on a common energy basis.
- **fatty acids**
  a major component of fats which are used by the body for energy and tissue development.
- **fauna**
  all the animal species that occur in a particular region.
- **favourable (conservation) status**
  the desirable status of a population for conservation purposes.
- **favourable condition**
  the desirable state for a site to be in for conservation purposes.
- **fazarabine**
  an investigational chemotherapy drug.
- **feasibility study**
  assessing whether a product is likely to make money.
- **fed cattle**
  slaughter cattle (usually steers and heifers) that have been finished on concentrated feed.
- **feed**
  animal feed bought in bulk by the group.
- **feed efficiency (feed conversion).**
  the units of feed consumed per unit of weight increase or unit of production (milk, meat, eggs).
- **feeder cattle**
  cattle, ready to be finished for market, weighing 550 to 650 pounds or heavier. Usu-

ally yearling cattle (between 1 and 2 years old) with a big frame.

- **feeder pigs**
young pigs, usually weighing 40 to 60 pounds, ready to be finished for market.

- **feedlot**
facility where cattle are confined in a small area and fed carefully mixed, high-concentrate feed to fatten them. On average, cattle are fed here about 5 months, although this can vary from 3 months to 9 months, depending on their weight when they enter and their desired final condition.

- **feedstock**
the raw material used for chemical or biological processes.

- **fen**
a MIRE that receives water from the surrounding land (unlike a BOG) and hence nutrients from rocks and soils. Because of this, a fen supports different communities of plants and animals than a BOG.

- **fenretinide**
a synthetic retinoid that may help prevent the development of some cancers.

- **fermentation**
the process of growing microorganisms for the production of various chemical or pharmaceutical compounds. Microbes are normally incubated under specific conditions in the presence of nutrients in large tanks called fermentors.

- **fertiliser**
a substance such as animal manure or an artificial chemical that is added to soil to increase its productivity for crops.

- **fertility**
the ability to conceive/be fertile.

- **fibre**
the cellulose portion of roughages (forages) that is low in TDN and hard to digest by monogastric animals.

- **fibrosis (fibrotic)**
  1. the growth of excessive amounts of fibrous tissue.
  2. of a condition marked by the presence of interstitial fibrous tissue, especially in the mammary gland resulting from mastitis.

Pulmonary Fibrosis

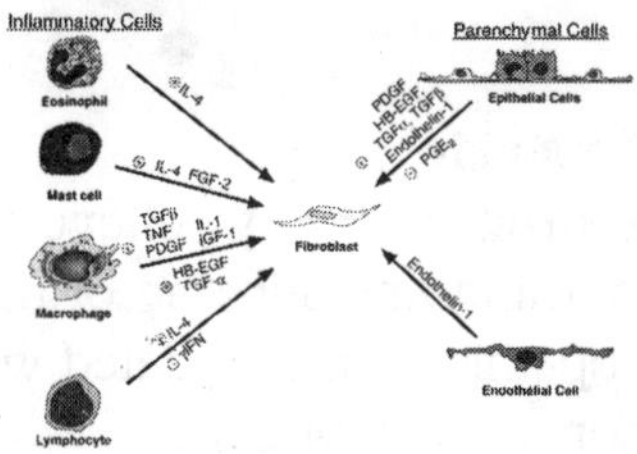

- **fill**
  a term designating the fullness of the digestive tract of an animal. Also used to identify a phase of packaging milk and/or milk products in which containers are filled with product.

- **filled milk**
  milk from which milk fat has been removed and replaced with other fats or oils.

- **filter DNA**
  a mastitis screening test useful in determining somatic cells present in milk by filtering and measuring the amount of deoxyribonucleic acid (DNA) produced.

- **fine-needle aspiration**
  use of a needle to remove fluid from a lump or cyst.

- **fingerlings**
  young, small fish. For trout, fingerlings are defined as fish from 1 to 6 inches long. For catfish, fingerling weight is 60 pounds per 1,000 fish and less.

- **finishing**
  the stage of production prior to slaughter where animals are typically fed rations of grain or other concentrates. The finishing stage is designed to increase an animal's weight and produce desirable carcass characteristics.

- **first-calf**
  a term commonly used to indicate the first calf born to bovine females.

- **fistula**
  an artificial opening into an animal, e.g. the rumen.

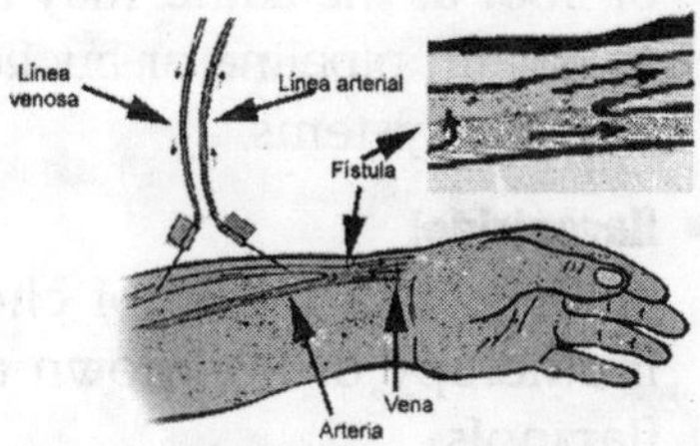

▪ **fixed dune**
a dune that has become stabilised by vegetation and is to a large extent, as a result, protected from further erosion by wind action.

▪ **flash pasteurisation**
a pasteurisation process which involves a high temperature, short-time treatment in which pourable products, such as juices, are heated for three to 15 seconds to a temperature that destroys harmful micro-organisms.

▪ **flat barn**
an area for milking cattle where the person milking is on the same level as the cow. May be used with a pipeline or bucket milking system. Generally the same area is used for cow housing.

▪ **flat barn milking area**
an area for milking cattle where the person milking cattle is on the same level or floor as the cattle. May be used with pipeline or bucket milking systems.

▪ **flavopiridol**
belongs to a group of chemotherapy drugs known as flavinols.

▪ **flavoured yogurt**
plain yogurt to which flavoring, but no fruit, has been added.

▪ **flecainide**
a drug that may relieve the burning and stinging of neuropathic pain associated with some types of cancer.

▪ **fledge**
to develop feathers.

▪ **fledgling**
a young bird that has grown feathers.

▪ **flip-chart**
paper holder to be used for training.

▪ **flora**
all the plant species that make up the vegetation of an area.

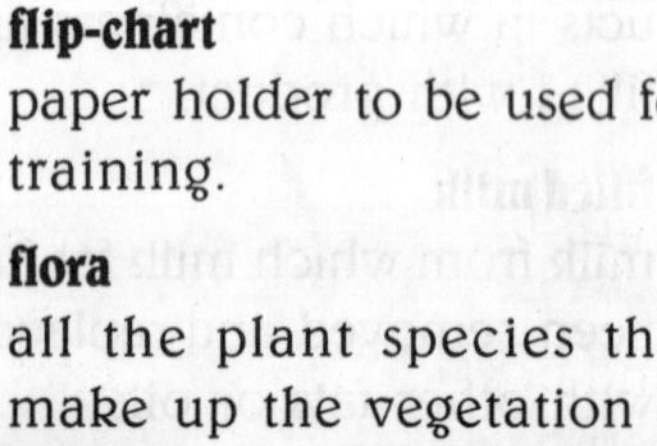

- **floxuridine**
  an anticancer drug that may stop the growth of tumour cells by interfering with the cells' ability to make DNA.

- **flt3l**
  a drug that increases the number of immune cells, and may stimulate the immune system to kill cancer cells.

- **fludarabine**
  a chemotherapy drug that interferes with tumour cell DNA.

- **fludeoxyglucose f 18**
  the radioactive form of glucose used in positron emission tomography (PET), a diagnostic imaging procedure.

- **fludrocortisone**
  a drug used to relieve symptoms of hormone shortage.

- **fluid milk**
  milk commonly marketed as fresh liquid milks and creams. It is the most perishable form of milk and commands the highest price per unit volume. Also called market milk.

- **fluid milk products**
  milk, flavoured milk, concentrated milk, filled milk, skim milk, fortified skim milk, lowfat milk, buttermilk, milk drinks and cream products.

- **fluoroscopy**
  an X-ray technique where live pictures appear on a screen instead of taking a permanent picture. Used while a procedure is being done to check on progress such as trying to basket a stone. With fluoroscopy it's possible to watch the interaction between the stone and the basket. ESWL uses fluoroscopy to aim the shock waves.

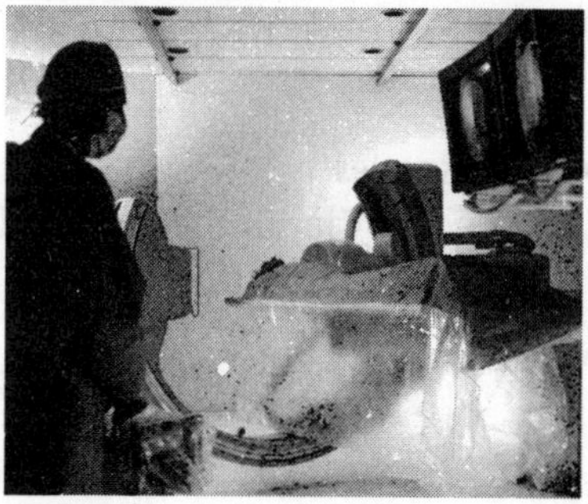

- **flush**
  a type of fen irrigated by a spring or soakway.

- **flush season**
  time during the year, usually spring and early summer, of maximum milk production.

- **flush system**
  a manure removal system in which an area is cleaned by high volumes of fresh water, or gray water that is recycled from a manure pit or lagoon.

- **foam**
  a gas dispersed in a liquid where the gas bubbles are the discrete phase. There are many food foams including whipped creams, ice cream, carbonated soft drinks, mousses, meringues, and the head of a beer. A foam is likewise unstable and needs a stabilising agent to form the gas bubble membrane.

- **focal species**
  a species that is the focus of conservation effort.

- **fodder**
  coarse food for cattle or horses, such as corn stalks or straw.

- **food allergy**
  a hypersensitive state resulting from the ingestion of a specific food antigen. Symptoms of a sensitivity can include allergic rhinitis, bronchial asthma, urticaria, angioneurotic edema, dermatitis, pruritus, headache, conjunctivitis, nausea, vomiting, diarrhea, colic, spastic constipation, mucous colitis, and perianal eczema. Food allergens are predominately protein in nature. The most common foods causing allergic reactions are wheat, milk, eggs, fish and other seafoods, chocolate, corn, nuts, strawberries, chicken, pork, legumes, tomatoes, cucumbers, garlic, and citrus foods. Foods that are rarely allergenic are rice, lamb, gelatin, peaches, pears, carrots, lettuce, artichokes, sesame oil, and apples. Diagnosis of a specific food allergy is obtained by a detailed food history, food diary, elimination diet, and cutaneous tests.

- **Food and Agriculture Organisation of the United Nations (FAO)**
  a UN organisation, founded in 1945, that collects and disseminates information about world agriculture. FAO also provides technical assistance to developing countries in agricultural production and distribution, food processing, nutrition, fisheries and forestry.

- **food intolerance**
  a person with a food intolerance is unable to digest and process that food correctly, usually due to a lack of a certain enzyme or enzymes. Intolerance can lead to allergy, however, if particles of undigested food manage to enter the bloodstream and cause a reaction.
- **food isolate**
  a microorganism that is derived from food for the purpose of identifying or characterising it.
- **food plant**
  a particular plant species favoured by a specific animal. For example, hare's-tail cotton-grass is the food plant of the large heath caterpillar.
- **food security**
  access by all people at all times to enough food for an active healthy life. Food security at a minimum includes the ready availability of nutritionally adequate and safe food, and an assured ability to acquire acceptable foods in socially acceptable ways, that is, without having to resort to emergency food supplies, scavenging, stealing, or other coping strategies. Causes of food insecurity may include poverty, civil conflict, governmental corruption, environmental degradation and natural disasters.
- **food thermometer**
  a special device that measures the internal temperature of cooked foods, such as meat, poultry, and any combination dishes to ensure that a safe food temperature is reached.

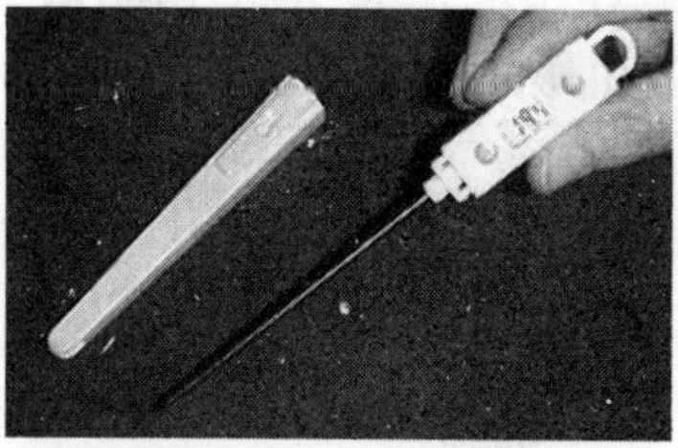

- **foodborne illnesses**
  illnesses caused by pathogens that enter the human body through foods.
- **foodborne outbreak**
  the occurrence of two or more people experiencing the same illness after eating the same food.
- **foodborne pathogens**
  disease-causing microorganisms found in food, usually bacteria, fungi, parasites, pro-

tozoans, and viruses. The top ten pathogens are: Salmonella; Staphylococcus Aureus; Campylobacter jejuni; Yersinia enerocolitica; Listeria monocytogenes; Vibro cholerae non-01; Vibrio Parahemolyticus; Bacillus cereus; Escherichia coli - enteropathogenic; and Shigella. Many of these pathogens may be found in contaminated meat, poultry, shell eggs, dairy products, and seafood.

- **foot and mouth disease (FMD)**

a highly contagious viral disease of cattle and swine, as well as sheep, goats, deer, and other cloven-hoofed ruminants. Although rarely transmissible to humans, FMD is devastating to livestock and has critical economic consequences with potentially severe losses in the production and marketing of meat and milk.

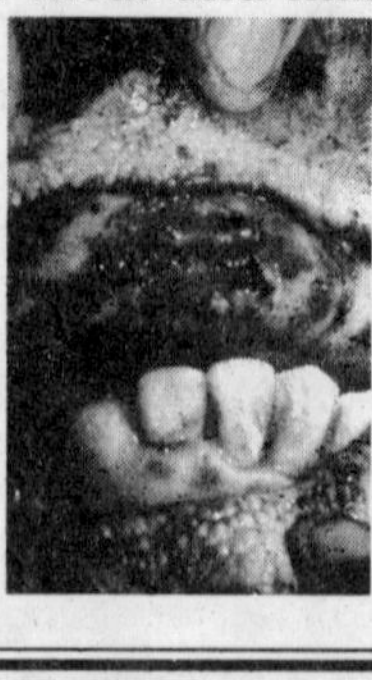

- **foot rot**

an inflammation occurring between the toes and in the hooves of sheep and cattle. It is caused by a combination of fungus and bacteria.

- **footbath**

a long shallow tub or depression in the concrete where cows walk through a mild solution (usually including copper sulfate or formalin) to promote foot health. Usually located along an alley where cows return from the milking parlour.

- **forage**

1. an animal's ability to search for grass.
2. feedstuffs composed primarily of the whole plant, including stems and leaves.
3. roughage of high feeding value. Grasses and legumes cut at the proper

stage of maturity and stored to preserve quality are forage. A crop that is high in fibre and grown especially to feed ruminant animals.

- **foraging**
behaviour associated with obtaining or consuming food.

- **foreign**
when this applies to the area of health it relates to something outside the normally accepted area.

- **forequarters**
the two front quarters of a cow. Also called the fore udder.

- **foreshore**
the lower zone of a beach extending between low-water spring- tide level and high- water spring- tide level.

- **forestripping**
expressing streams of milk from the teat prior to machine milking to determine visual quality and to stimulate 'milk letdown'.

- **forewing**
one of the pair of wings nearest the head in insects such as butterflies, moths and dragonflies. Compare with hindwing.

- **forming groups**
setting up groups of milk producers.

- **fortified**
the addition of nutrients not initially present or present in insignificant amounts in foods. Dairy products are often fortified with vitamin A and vitamin D.

- **founder (laminitis)**
inflammation of foot and lower leg of ruminant animals caused by overeating grain or green grass.

- **free range or free roaming**
livestock or poultry has been allowed access to the outside.

- **free stalls**
resting cubicles or 'beds' in which dairy animals are free to enter and leave, as opposed to being confined in stanchions.

- **free-choice**
a feeding system that allows animals to eat at will. See **self-feeder**.

freemartin
female born twin to a bull calf (about 9 out of 10 of these will not conceive). A sterile heifer born twin with a bull.

freestalls
resting cubicles or 'beds' in which dairy cows are free to enter and leave, as opposed to being confined in stanchions or pens.

freeze branding
a method of identification with a cold iron to aid in easily identifying dairy cattle.

freeze drying
the evaporation of water from a frozen product with the aid of high vacuum. Also called lyophilisation.

french size
many medical tubes and catheters are measured by French Size. This represents the circumference in mm of the tube. The higher the number, the larger the tube.

fresh
a cow that has recently given birth to a calf. Also refers to dairy products having original qualities unimpaired and those recently produced or processed.

fresh (poultry)
poultry whose internal temperature has never been below 26°F.

fresh cow
a cow that has recently given birth to a calf.

fresh ham
the uncured leg of pork. Since the meat is not cured or smoked, it has the flavour of a fresh pork loin roast or pork chops. Its raw colour is pinkish red and after cooking, greyish white.

fresh milk
dairy products having original qualities unimpaired and those recently produced or processed.

freshen
commonly used to designate the act of calving (parturition), to give birth to a calf and concurrently initiate lactation.

fromage frais
a smooth, soft fresh cheese used to make Petit Miam and Frûche dairy foods.

fryer-roaster turkey

a young immature turkey, usually under 16 weeks of age, of either sex, that is tender-meated with soft, pliable, smooth-textured skin, and flexible breastbone cartilage.

full-sibs

animals having the same sire and dam, such as full brothers, full sisters, and/or a full brother-sister pair.

fully cooked

fully cooked product needs no further cooking. The product is fully cooked in the plant, and it can be reheated or eaten directly from the package. Also known as ready-to-eat.

functional foods

food that affects beneficially one or more target functions in the body, beyond adequate nutritional effects in a way which is relevant to either an improved state of health and well-being, or reduction of risk of disease.

functional genomics

a field of research that aims to understand what each gene does, how it is regulated and how it interacts with other genes.

further processing

smoking, cooking, canning, curing, refining or rendering in an official establishment of product previously prepared in official establishments.

fusion

joining of the membrane of two cells, thus creating a daughter cell that contains some of the same properties from each parent cells. Used in making hybridomas.

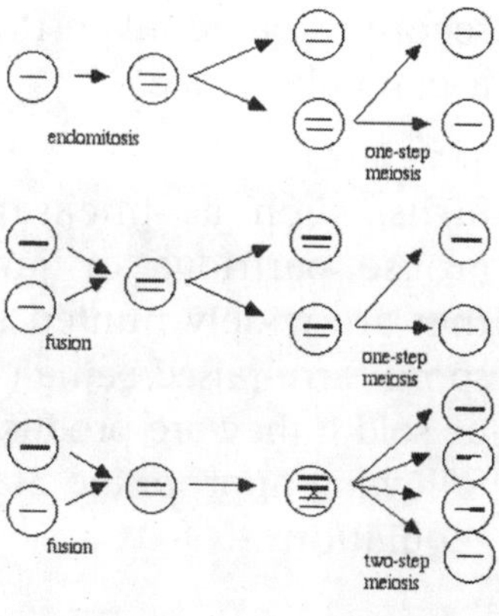

- **fyke net**
  a net used to catch eels.
- **g1147211**
  belongs to the group of drugs known as carnptothecin analogues.
- **gait**
  manner of walking.
- **galactopoietic**
  stimulating or increasing secretion of milk.
- **gallium nitrate**
  a drug that lowers blood calcium. Used as treatment for hypercalcemia (too much calcium in the blood) and for cancer that has spread to the bone (bone metastases).
- **game**
  mammals, birds or fish hunted for sport and/or food.
- **game bag**
  record of game taken from a particular place.
- **game birds**
  birds, such as pheasant, grouse, partridge, or quail, that are widely hunted for sport. Farm-raised game can be sold if they are produced following appropriate State regulations.

- **ganciclovir**
  a drug that interferes with DNA synthesis and stops the growth of cancer cells.
- **garget**
  a common term for an inflammation of the udder of the cow or the resulting abnormal milk. More accurately referred to as mastitis.
- **gastroenteritis**
  chemical, bacterial, or viral inflammation of the mucosa of the stomach and intestines.
- **gastrointestinal**
  the stomach and intestine.
- **gel electrophoresis**
  a process for separating molecules by forcing them to migrate through a gel

under the influence of an electric field.

- **gelatin**
thickener from collagen which is derived from the skin, tendons, ligaments, or bones of livestock. It may be used in canned hams or jellied meat products, as well as non-food products such as photography and medicine.

- **gelatine**
a water soluble protein that functions as a gelling agent to gel or set firm the texture of foods.

- **gem 231**
an investigational drug that may inhibit the growth of malignant tumours.

- **gemcitabine**
a drug that stops tumour growth by disrupting the DNA in cancer cells. Also known as Gemzar

- **gene**
  1. a part of a chromosome in the cells of an organism that controls the inheritance of a particular trait or characteristic. For example, in humans there are genes that control eye colour.
  2. a segment of chromosome. Some genes direct the syntheses of proteins, while others have regulatory functions. See **Operator gene**, **Structural gene**, **Suppressor gene**.
  3. a unit of DNA that determines and transmits hereditary characteristics from parent to offspring
  4. the basic unit of heredity, an ordered sequence of nucleotide bases in a segment of DNA.

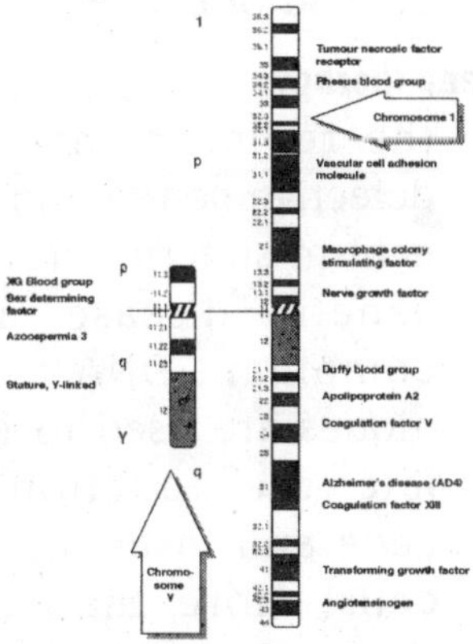

- **gene amplification**
the increase, within a cell, of the number of copies of a given gene.

- **gene knockout**
the replacement of a normal gene with a mutated form of the gene by using homolo-

gous recombination. Used to study gene function.

- **gene machine**
  a computerised device for synthesising genes by combing nucleotides (bases) in the proper order.

- **gene mapping**
  determination of the relative locations of genes on a chromosome.

- **gene sequencing**
  determination of the sequence of nucleotide bases in a strand of DNA.

- **gene therapy**
  1. the replacement of a defective gene in an organism suffering from a genetic disease. Recombinant DNA techniques are used to isolate the functioning gene and insert it into cells. More than 300 single-gene genetic disorders have been identified in humans. A significant percentage of these may be amenable to gene therapy.
  2. treatment to modify the genes of a person's white blood cells or tumour cells to try to control or cure the cancer.

- **gene-modified**
  cells that have been altered to contain different genetic material than they originally did.

- **genetic**
  relating to genes or genetics.

- **genetic base**
  the average genetic merit of a population (usually a breed) at a specific period, which is used as a reference point to express a genetic difference from a base population (genetic merit of cows and bulls).

- **genetic code**
  1. the code by which genetic information in DNA is translated into biological function. A set of three nucleotides (codons), the building blocks of DNA, signifies one amino acid, the building blocks of proteins.
  2. the biochemical basis of heredity; the sets of nucleotide bases in a DNA sequence that specify each amino acid in the sequence of a protein.

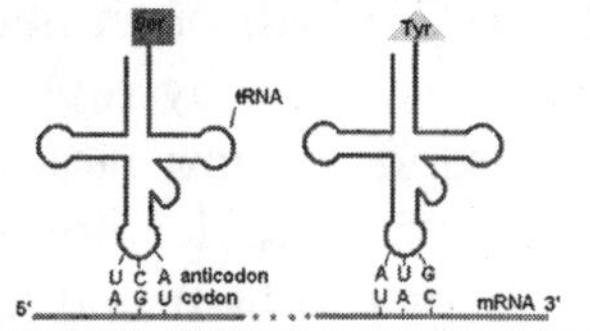

2nd base in codon

1st base in codon	U	C	A	G	3rd base in codon
U	Phe Phe Leu Leu	Ser Ser Ser Ser	Tyr Tyr STOP STOP	Cys Cys STOP Trp	U C A G
C	Leu Leu Leu Leu	Pro Pro Pro Pro	His His Gln Gln	Arg Arg Arg Arg	U C A G
A	Ile Ile Ile Met	Thr Thr Thr Thr	Asn Asn Lys Lys	Ser Ser Arg Arg	U C A G
G	Val Val Val Val	Ala Ala Ala Ala	Asp Asp Glu Glu	Gly Gly Gly Gly	U C A G

The Genetic Code

- **genetic defects**
  inherited traits which can cause problems.

- **genetic gain**
  the result of breeding increasingly superior bulls to increasingly superior cows and can be measured by the increased milk production of these cows year on year).

- **genetic merit**
  the genetic value of the animal used in a breeding program. See **breeding value**.

- **genetic modification**
  a number of techniques, such as selective breeding, mutagenesis, transposon insertions and recombinant DNA technology, that are used to alter the genetic material of cells in order to make them capable of producing new substances, performing new functions or blocking the production of substances.

- **genetic predisposition**
  susceptibility to disease that is related to a genetic predisposition mutation, which may or may not result in actual development of the disease.

- **genetic screening**
  the use of a specific biological test to screen for inherited diseases or medical conditions. Testing can be conducted prenatally to check for metabolic defects and congenital disorders in the developing foetus as well as postnatally to screen for carriers of heritable diseases.

- **genetic testing**
  the analysis of an individual's genetic material. Genetic testing can be used to gather information on an individual's genetic predisposition to a particular health condition, or to confirm a diagnosis of genetic disease.

- **genetic trend**
  genetic change per year for a trait in the population.

■ **genetics**
the study of genes and of inheritance and variation in organisms.

■ **genome**
1. the total hereditary material of a cell, comprising the entire chromosomal set found in each nucleus of a given species.
2. the total hereditary composition of a cell or organism.

Duplications dans le génome d '*Arabidopsis thaliana*
Chaque duplication est numérotée et symbolisée par une couleur. ex: I-IV A entre les chromosomes I et IV

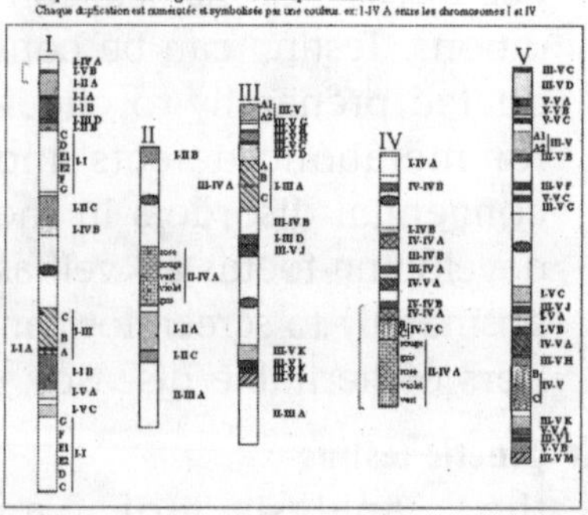

■ **genomics**
the study of genes and their function. Recent advances in genomics are bringing about a revolution in our understanding of the molecular mechanisms of disease, including the complex interplay of genetic and environmental factors. Genomics is also stimulating the discovery of breakthrough health-care products by revealing thousands of new biological targets for the development of drugs and by giving scientists innovative ways to design new drugs, vaccines and DNA diagnostics. Genomic-based therapeutics may include 'traditional' small chemical drugs, protein drugs and gene therapy.

■ **genotype**
1. genetic makeup of an individual or group. Compare Phenotype.
2. the actual genetic constitution (makeup) of an individual as determined by its germ plasm. For example, there are two genotypes for black-and-white Holsteins—BB and Bb.

■ **genus**
a unit used in the classification of plants and animals. It consists of a grouping of closely related species. When the scientific name for a particular species is given the name of the genus (the 'generic name') is given first e.g. the lady's slipper orchid -Cypripedium calceolus -

belongs to the genus called 'cypripedium'.

- **geology**
  the study of the Earth's crust and its rocks.

- **gerber test**
  test to determine amount of fat in milk.

- **germ cell**
  reproductive cell (sperm or egg). Also called gamete or sex cell.

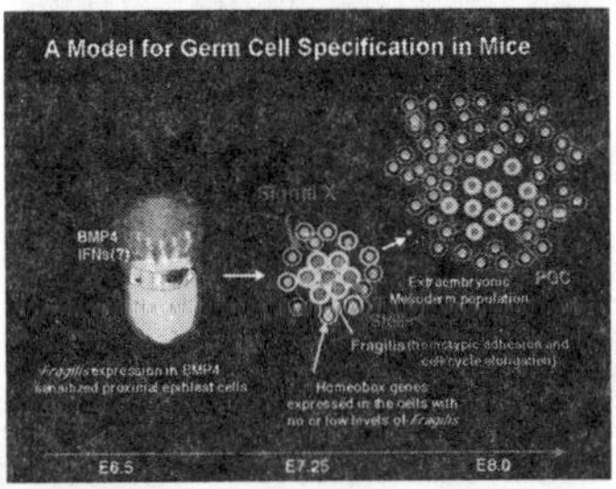

- **germicidal**
  a substance that has the ability to kill germs.

- **germplasm**
  the total genetic variability, represented by germ cells or seeds, available to a particular population of organisms.

- **gestation pregnancy (gravidity).**
  the period from conception to birth. The period of foetal development between fertilisation of the ovum and birth of the offspring.

- **get the offspring of a male.**
  a get-of-sire refers to given number (commonly four) of progeny from a sire.

- **ghee**
  concentrated milk fat prepared by melting butter, decanting the fat after gravity separation from the serum, and driving off most of the remaining moisture by heating. About 1 percent moisture remains. It is used mostly in Asia and Africa.

- **giblets**
  giblets are the heart, liver and gizzard of a poultry carcass. Although often packaged with them, the neck of the bird is not a giblet.

- **gilt**
  female hog, usually less than 15 months of age, which has not produced a litter.

- **giving milk**
lactating, or the act of yielding milk by a mammal.

- **gizzard**
the gizzard is the mechanical 'stomach' of a bird. It is located just after the true or glandular stomach in the gastrointestinal system. Since poultry have no teeth and swallow feed whole, this muscular organ, sometimes called 'hen's teeth', mechanically grinds and mixes the bird's feed.

- **glial tumours**
a general term for many types of tumours of the central nervous system, including astrocytomas, ependymal tumours, glioblastoma multifonne and primitive neuroectodertnal tumours.

- **glioblastoma**
a general term that refers to malignant astrocytoma.

- **glioma**
a general term for many types of cancers of the brain and spinal cord.

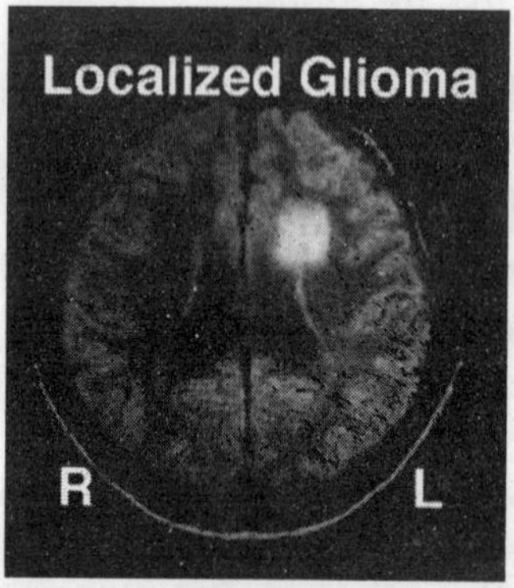

- **gliosarcoma**
a type of glioma.

- **globally threatened**
a category of threat that indicates that a taxon is threatened on a world scale.

- **globulin**
group of proteins that are soluble in salt solution, but insoluble in water.

- **gluten**
the insoluble protein constituent of wheat and other grains. It is obtained from flour by washing out the starch and is used as an adhesive agent, giving dough its tough, elastic character. See **Celiac Disease**.

- **gluten-free diet**
  the elimination of gluten-containing grains and products used in the treatment of various medical conditions such as Celiac Disease. Eliminated non-acceptable gluten-containing grains and foods include: Barley, Bulgur, Couscous, Durum, Einkorn, Emmer, Farina, Graham, Groats, Kamut, Mir, Oats, Triticale, Rye, Semolina, Spelt, and Wheat (bran, germ, flour, starch). Acceptable grains and starches may include: Amaranth, Arrowroot, Artichoke, Buckwheat (kasha), Cassava (Tapioca, Manioc), Chickpea (Garbanzo Bean), Corn, Fava or Faba Bean, Grits (corn), Hemp, Hominy (corn), Nut flours (almond, chestnut, hazelnut, etc.), Milo (Sorghum), Millet, Pea, Bean, Mung Bean, Lentils, Peanuts, Potato, Quinoa, Rice, Sago, Sesame, Soy, Sunflower, Sweet Potato, Sweet Rice flour, Tef and Wild Rice.

- **gluten-free, casein-free diet**
  a diet that is free from grains that contain gluten, as well as, dairy foods or products that contain the protein casein.

- **glycoprotein**
  a protein conjugated with a carbohydrate group.

- **Goat's milk**
  contains about the same protein, fat, and carbohydrate content as cow's milk. However, goat's milk and cow's milk differ with regard to a number of nutrients — cow's milk is available in lower fat versions, while goat's milk is sold only as whole milk; goat's milk is lower in vitamin B12 than cow's milk and it is not fortified with vitamin D.

- **goitrogenic**
  producing or tending to produce goiter (enlargement of the thyroid gland).

- **good manufacturing practices (GMP)**
  the quality system regulation that includes requirements related to the methods used in, and the facilities and controls used for, designing, manufacturing, packaging, labelling, installing or servicing devices.

- **gossypol**
  a toxic yellow pigment found in cottonseed. Heat and pressure tend to bind it with protein and thereby render it safe for animal consumption.

- **gout**
  a disease caused by excessively high uric acid in the blood. This high uric acid blood level will sometimes allow formation of uric acid crystals in the joints, especially at the base of the large toe. This causes intense joint pain. Can contribute to calcium and uric acid stone disease (See **Gouty Diathesis**). Usually treated with allopurinol when associated with kidney stones.

- **gouty diathesis**
  condition in which gout is associated with uric acid or calcium oxalate kidney stone production, usually from extremely acid urine or elevated urinary uric acid excretion.

- **GPD (gaseous products of digestion)**
  these include the combustible gases produced in the digestive tract during fermentation of the ration. Methane constitutes the major proportion of the combustible gases produced by ruminants; however, non-ruminants also produce methane. Trace amounts of hydrogen, carbon monoxide, acetone, ethane, and hydrogen sulphide are also produced.

- **grade**
  1. the grade of a tumour is determined by how different the tumour cells are from normal cells, the growth rate of the tumour, and its tendency to spread (infiltrate). The systems used to grade tumours vary with each type of cancer.
  2. animals showing the predominant characteristics of a given breed. They usually have at least one purebred parent, ordinarily the bull.

- **grade a eggs**
  cover a relatively small area, have a yolk that is round and upstanding and contain a large proportion of thick white to thin white.

- **grade A milk**
  milk produced and processed under rigid sanitary regulations. Milking and processing facilities are approved and inspected by public health officials. Fluid milk products must meet this inspection standard.

- **grade AA eggs**
  stand up tall, have a firm yolk which covers only a small area, and have a large proportion of thick white to thin white.

- **grade B eggs**
  more spread out than Grade A eggs, have a flat yolk and contain as much (or more) thin white as thick white.

- **grade C eggs**
  cover a very wide area in which most of the white is thin and the yolk is very flat.

- **grade cattle**
  an animal possessing the distinct characteristic of a particular breed but not registered with a breed association. Sometimes called commercial cattle.

- **grade butter**
  grading system based primarily on flavor and secondarily on body, color, and salt characteristics. There are three grades. AA is the grade typically sold in supermarkets.

- **grade eggs**
  grading system based on egg's appearance when it is broken. There is no difference in nutritive value between the different grades. See **Grade AA Eggs**, **Grade C Eggs**.

- **grading up**
  the continued use of purebred sires on grade dams.

- **graft-versus-host disease**
  an immune response to the body's normal tissue by immune cells present in a donor's transplanted tissue, such as bone marrow or peripheral stem cells.

- **graft-versus-tumour**
  an immune response to a person's tumour cells by immune cells present in a donor's transplanted tissue, such as bone marrow or peripheral blood.

- **granulocyte**
  one of three types of white blood cells. Granulocytes digest bacteria and other parasites.

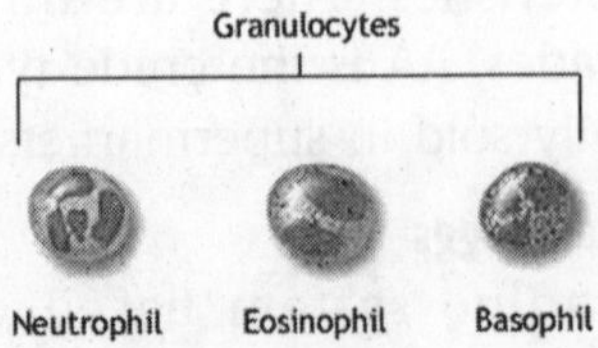

- **granulocyte-macrophage colony stimulating factor (GMCSF)**
  a natural hormone that stimulates white blood cell production, particularly that of granulocytes and monocytes (the precursors of macrophages).

- **grass tetany**
  a magnesium-deficiency disease of cattle characterised by hyperirritability, muscular spasms of legs and convulsions.

- **grey water**
  water that is considered waste and not to be used for cleaning milking systems. Usually including recycled water from a lagoon or milk house waste. Even water only used to cool milk in a plate cooler is considered gray water, though it is often fed to cows to reduce total usage.

- **graze**
  to consume standing vegetation, as by livestock or wild animals.

- **green cheese**
  a term applied to the fresh uncured product. It refers to flavour, odour, body and texture, not colour.

- **green chop (fresh forage)**
  forages harvested (cut and chopped) in the field and hauled to livestock. This minimises the loss of moisture, colour, nutrients and wastage. Also called zero grazing or soilage.

- **greenhouse barn**
  commonly referring to any of a group of hoop type barns with translucent or plastic covers over a tubular steel frame.

- **grike**
  a cleft or fissure created by the action of water along a line of weakness in limestone rock.

- **grip**
  a moorland drainage ditch.

- **grooved concrete**
  floor surfaces with grooved patterns cut or depressed

into concrete to provide better traction for cattle.

- **gross energy (GE)**
  the amount of heat, measured in calories, released when a substance is completely oxidised in a bomb calorimeter.

- **ground beef**
  'ground beef' or 'chopped beef' shall consist of chopped fresh and/or frozen beef with or without seasoning and without the addition of beef fat as such, shall not contain more than 30 percent fat, and shall not contain added water, phosphates, binders or extenders.

- **ground game**
  rabbits and hares.

- **growth factors**
  1. naturally occurring proteins that stimulate the growth and reproduction of specific cell types. Growth factors are essential to regenerative medicine and tissue engineering.
  2. substances made by the body that function to regulate cell division and cell survival. Some growth factors are also produced in the laboratory and used in biological therapy. Growth factor antagonists, such as CEP-2563 dihydrochloride, stop tumour cells from growing

- **growth hormone**
  a protein produced by the pituitary gland that is involved in cell growth. Human growth hormone is used clinically to treat dwarfism. Various animal growth hormones can be used to improve milk production as well as produce a leaner variety of meat. Also called somatotropin.

- **growth pattern**
  the characteristic way in which a particular organism grows.

- **growthy**
  describes an animal that is large and well-developed for its age.

- **groyne**
  an artificial construction built out into the sea in order to keep material on a beach.
- **guar gum**
  a water-soluble paste made from the seeds of the guar plant and used as a thickener and stabiliser in foods and pharmaceuticals. May be substituted for xanthan gum in most recipes.
- **gurney**
  a mobile bed with wheels designed for transport of patients in hospitals and ambulances.
- **gutter**
  a shallow to deep channel located behind cows in tiestall barns capture manure and urine.
- **gynaecologic cancer**
  cancer of the female reproductive tract, including the cervix, endometrium, ovaries, uterus and vagina.
- **H bonding**
  some occurs between the individual caseins in the micelle but not much because there is no secondary structure in casein proteins.
- **habitat**
  the place in which an animal or plant lives.
- **hair tube**
  a device/method for surveying for certain mammal species that relies upon animals leaving hair on the sticky surfaces of tubes that are baited with food or left on commonly-used pathways.
- **hairy cell leukemia**
  a rare type of chronic leukemia in which the abnormal white blood cells appear to be covered with tiny hairs.
- **halal and zabiah halal**
  products prepared by federally inspected meat packing plants identified with labels bearing references to 'Halal' or 'Zabiah Halal' must be handled according to Islamic law and under Islamic authority.
- **half and half**
  a mixture of milk and cream which cannot legally be called cream because its percent of milk fat is too low (between 10.5% and 18%). It is almost always homogenised.

- **half-sib**

in genetics, a half brother or half sister.

- **ham**

ham means pork which comes from the hind leg of a hog. Ham made from the front leg of a hog will be labelled 'pork shoulder picnic'. Hams may be fresh, cured, or cured-and-smoked. The usual colour for cured ham is deep rose or pink, fresh ham (which is not cured) has the pale pink or beige colour of fresh pork roast, country hams and prosciutto (which are dry cured) range from pink to mahogany colour. 'Turkey ham' must be made from the thigh meat of turkey.

- **ham steak**

another name for centre cut ham slices.

- **hamburger**

'hamburger' shall consist of chopped fresh and/or frozen beef with or without the addition of beef fat as such and/or seasoning, shall not contain more than 30 percent fat, and shall not contain added water, phosphates, binders, or extenders.

- **hand milking**

1. the manual milking of an animal as opposed to the use of mechanical milking devices.
2. the manual milking of an animal as opposed to mechanical milking.

- **handlers**

processors or dealers of milk who commonly purchase raw milk and sell pasteurised milk and milk products.

- **haploid**

a cell with half the usual number of chromosomes, or only one chromosome set. Sex cells are haploid. Compare Diploid.

- **hapten**

the portion of an antigen that determines its immunological specificity. When coupled to a large protein, a hapten stimulates the formation of antibodies to the two-molecule complex. Also called antigenic determinant.

- **hardware disease**

commonly, an inflammation of the body cavity by an animal eating metal objects and perforation of the digestive tract.

- **harrow**
  to break up clods of earth on heavy ground.
- **harvestore silo**
  a brand of oxygen limiting (air tight) upright silos with bottom unloading.
- **hay**
  1. grass cut during spring and stored (and dried) in bails. Usually fed during winter, when cows are not being milked and when grass supply is decreased to maintain their body weight.
  2. dried feed consisting of the entire plant. Alfalfa, clover, grass, and oat hay are used in dairy rations.
- **hay dried forage (e.g., grasses, alfalfa, clovers)**
  used for feeding farm animals.
- **hay meadow**
  an enclosed grassland which is cut two or three times in the summer to produce a crop of hay to be fed to livestock.
- **haylage**
  low-moisture silage (35 to 55 percent moisture). Gass and legume crops are cut and wilted in the field to a lower moisture level than normal for grass silage, but the crop is not sufficiently dry for baling. It is commonly stored in a sealed, or air-tight, storage system.
- **hazard**
  a biological, physical, or chemical property that may cause a food to be unsafe for human consumption.
- **headcheese**
  a jellied product consisting predominantly of pork by-products and seasoning ingredients. It must contain some product from the head.
- **headland**
  on farms, headlands are small areas of semi-natural habitat (e.g. rough grassland) that extend into cropped areas.
- **headlocks**
  self-locking stanchions along a feed alley in which cows voluntarily enter the head slot when going to eat. All cows may be held until herd health work is completed, and then all cows

may be simultaneously released. Headlocks may be adjusted to remain open, allowing cows to come an go at will, when restraining the cows is not necessary.

- **heart girth**

  the circumference of the body just back of the shoulders of an animal. It is used to estimate body weight.

- **heat**

  at temperatures above the boiling point casein micelles will irreversibly aggregate. On heating, the buffer capacity of milk salts change, carbon dioxide is released, organic acids are produced, and tricalcium phophate and casein phosphate may be precipitated with the release of hydrogen ions.

- **heat increment (HI)**

  the increase in heat produced following consumption of food. It consists of calories released in fermentation and nutrient metabolism. When environmental temperature is below critical temperature, this heat may be used to keep the body warm, therefore, it is not wasted. Also called work of digestion.

- **heat period**

  that period of time when a female will accept a male in the act of mating. Also called in heat or estrus.

- **heat treatment**

  destroying any potential pathogenic germs by heating milk at a minimum of 63°C for 30 min.

- **heat treatment**

  whey proteins become adsorbed, altering the behaviour of the micelle.

- **hectare**

  a metric unit of land measurement (2.47 acres).

- **heifer**

  1. a bovine female that has not given birth to a calf.
  2. a bovine female less than three years of age who has not borne a calf. Young cows with their first calves are often called first-calf heifers.

- **helicobacter pylori**
  bacteria that may cause inflammation of the stomach. They are found in persons with chronic gastritis, ulcers or lymphoma of the stomach.
- **hemagglutination**
  clumping (agglutination) of red blood cells.
- **hemangiopericytoma**
  a form of cancer involving blood vessels and soft tissue.
- **hematologic malignancies**
  cancer of the blood or bone marrow, including leukemia and lymphoma. Also called hematologic cancer.
- **hen**
  an adult female chicken or turkey.

- **hen, fowl (baking or stewing)**
  a bird of this class is a mature female chicken, usually more than 10 months of age, with meat less tender than that of a roaster or roasting chicken and nonflexible breastbone tip.
- **heparin**
  a drug that helps prevent blood clots from forming.
- **HER2/neu**
  the gene that controls cell growth by making the human epidermal growth factor receptor. Also known as c-erbB-2. The protein produced is called the HER2/neu antigen.
- **herb**
  any non-woody vascular plant.
- **herbicide**
  a chemical that kills plants or suppresses plant growth.
- **herbivore**
  a plant-eater.

- **herbivorous**
  plant-eating.
- **herd**
  a group of animals (especially cattle), collectively considered as a unit.

- **herd average**
average milk and component production per cow for all cows in the herd unit for 12 months.

- **herd improvement**
various measures which contribute to increased production from commercial dairy herds.

- **herd owner**
the declared owner of a herd of dairy cattle as recorded on a legal document, registration papers or other official documents.

- **herd recording**
collection of statistics which relate to various aspects of milk production (e.g. mating, milk production, disease).

- **herd tests**
the milk samples which are taken periodically to tell a farmer how good an individual cow is.

- **herdmate comparison**
method of estimating the genetic transmitting ability in which a cow's yield is compared to that of other cows in the same breed from different sires that calved in the same herd during the same season.

- **heredity**
transfer of genetic information from parent cells to progeny.

- **heritability**
  1. the degree to which the phenotype (expression of a trait that you can measure, or describe like how much milk, or what colour hair) of an animal is likely to predict the expression of that trait in the offspring of the animal. How likely is it for a cow that produces a lot of milk to have a daughter that produces a lot of milk?
  2. the proportion (fraction) of difference among animals for a trait, such as milk production, due to genetic difference rather than environmental factors. The frac-

tion of variation in a trait that is genetically transmissible from parent to offspring.

- **herringbone milking parlour**
  a raised, zigzag-designed milking stall which allows the group milking of several cows at one time in a pipeline milking system.

- **herringbone parlour**
  a milking parlour in which cows stand side-by-side, angled towards the pit. This allows milking from the side of the udder.

- **hibernation**
  a period of inactivity that occurs in some mammals and other vertebrates during winter. It is an adaptation to conserve energy when there is little food available.

- **hickory-smoked ham**
  a cured ham which has been smoked by hanging over burning hickory wood chips in a smokehouse. The ham may not be labelled 'hickory smoked' unless hickory wood has been used.

- **high forest**
  woodland managed to allow the majority of trees to reach maturity. Compare with coppice.

- **high-grade lymphomas**
  includes large cell, immunoblastic, lymphoblastic, and small noncleaved cell lymphomas. These lymphomas have a high growth rate and a lower response rate to most chemotherapy regimens than that seen with intermediate grade lymphomas.

- **high-moisture silage**
  silage usually containing 70 percent or more moisture.

- **hindwing**
  one of the pair of wings behind those nearest the head (forewing) in insects such as butterflies, moths and dragonflies.

- **hinterland**
  the region lying inland of a coastline.

- **histocompatibility**
  immunologic similarity of tissues such that grafting can be done without tissue rejection.

- **histocompatibility antigen**
  an antigen that causes the rejection of grafted material

from an animal different in genotype from the host animal.

- **hodgkin's disease**
a malignant disease of the lymph nodes that is characterised by painless enlargement of lymphatic tissues and the spleen. It often involves symptoms such as fever, weight loss, anaemia, and night sweats.

- **holding area**
an area to hold cows prior to entry into the milking parlour. Sometimes called a holding pen. May also include a wash pen.

- **holding pen**
an area in which cows congregate prior to entering a milking parlour to be milked.

- **homeobox**
family of genes that regulates activities of other genes (turns genes on and off).

- **homeopathic remedies**
small doses of medicines and herbs that are believed to stimulate the immune system by causing the same symptoms in healthy people as those of the disease being treated

- **homogenisation**
process to stop the gravity separation of the fat in milk to form a cream layer by breaking down the size of the fat globules to create a stable fat emulsion. All Pure milk is homogenised.

- **homogenisation**
process of breaking up the fat globules of whole milk into extremely small particles so that the fat remains emulsified and does not rise to the top to form cream.

- **homogenised**
process that ensures that milk has a uniform flavour and appearance by breaking up milk-fat and spreading it finely throughout the milk. This is done in a homogenise where, under pressure, the milk is forced through tiny openings.

- **homogenised milk**
milk that has been treated to ensure breakup of fat globules to such an extent that, after 48 hours of quiescent storage at 7°C, no visible cream separation occurs on the milk. The re-

duced size of fat particles results in formation of a softer curd in the stomach.

- **homoharringtonine**
a drug used to stop tumour cells from dividing.

- **homologous**
corresponding or alike in structure, position or origin.

- **honey-cured**
honey-cured may be shown on the labelling of a cured product if honey is the only sweetening ingredient or is at least half the sweetening ingredients used and if the honey is used in an amount sufficient to flavour and/or affect the appearance of the finished product.

- **honeydew**
a sugar- rich substance exuded by APHIDS and some other sap-sucking BUGS.

- **hormone**
a chemical or protein that acts as a messenger or stimulatory signal, relaying instructions to stop or start certain physiological activities. Hormones are synthesised in one type of cell and then released to direct the function of other cell types.

- **hormone therapy**
treatment of cancer by removing, blocking or adding hormones. Also called endocrine therapy.

- **hormones**
chemicals produced by glands in the body. Hormones control the actions of certain cells or organs.

- **host**
a cell or organism used for growth of a virus, plasmid or other form of foreign DNA, or for the production of cloned substances.

- **host-vector system**
combination of DNA-receiving cells (host) and DNA-transporting substance (vector) used for introducing foreign DNA into a cell.

- **hot quarter**
a quarter of the udder that is infected and may actually feel hard or hot to the touch due to elevated temperatures.

- **hotspot**
an area with a high concentration of biodiversity.

- **hoverfly**
two-winged (or 'true') flies of the family 'syrphidae'.

The adults are often brightly coloured insects with great flying and hovering ability. See **diptera**.

- **HTST (high-temperature short-time)**
the temperature and time combination for pasteurisation of milk and milk products. The minimum for milk is 71.7°C (161°F) for at least 15 seconds. Higher temperatures are used in pasteurising products of high solids because of the protection afforded bacteria by solids.

- **human genome project**
an international research effort aimed at discovering the full sequence of bases in the human genome. Led in the United States by the National Institutes of Health and the Department of Energy.

- **human immunodeficiency virus (HIV)**
the virus that causes acquired immune deficiency syndrome (AIDS).

- **humectant**
a substance added to foods to help retain moisture and soft texture. An example is glycerine, which may be used in dried meat snacks.

- **hundredweight (cwt.)**
generally refers to milk sold off the farm in 100 pound units.

- **hutch**
an individual housing unit for young calves. Often made of white fibreglass or polyvinyl.

- **hybrid**
an organism that is produced from a cross between parents from two distinct species.

- **hybrid vigour (heterosis)**
the increased growth or production exhibited by the first generation over the average of its two parent breeds. It is usually expressed as a percentage.

- **hybridisation**
production of offspring, or hybrids, from genetically dissimilar parents. The process can be used to produce hybrid plants (by crossbreeding two different varieties) or hybridomas (hybrid cells formed by fusing two unlike cells, used in producing monoclonal antibodies).

- **hybridoma**
the cell produced by fusing two cells of different origin. In monoclonal antibody technology, hybridomas are formed by fusing an immortal cell (one that divides continuously) and an antibody-producing cell. See **Monoclonal antibody**.

- **hydraulic connection**
connected by shared groundwater.

- **hydrocortisone**
a drug used to relieve the symptoms of certain hormone shortages and to suppress an immune response.

- **hydrolysed (source) protein**
flavour enhancers that can be used in meat and poultry products. They are made from protein obtained from a plant source such as soy or wheat, or from an animal source, such as milk. The source used must be identified on the label.

- **hydromorphone**
a drug used to relieve pain. Also called Dilaudid.

- **hydrophobic interactions**
caseins are among the most hydrophobic proteins and there is some evidence to suggest they play a role in the stability of the micelle. It must be remembered that hydrophobic interactions are very temperature sensitive.

- **hydroxyurea**
belongs to the group of chemotherapy drugs known as DNA synthesis inhibitors.

- **hymenoptera**
an insect order that includes ants, bees and wasps.

- **hypercalcemia**
  1. above normal blood calcium levels. In kidney stone disease, this frequently is due to hyperparathyroidism.
  2. high levels of calcium in the blood.

- **hypercalciuria**
  1. excessive urinary calcium.
  2. abnormally high urine oxalate levels.

- **hyperparathyroidism**
  uncontrolled, excessive secretion of the parathyroid , lands, producing a disease characterised by loss of calcium from the bones. Often resulting in high serum calcium levels and kidney stones.

- **hyperplasia**
  an abnormal increase in the number of cells in an organ or tissue.

- **hypersensitivity**
  an exaggerated response by the immune system to a substance or drug.

- **hyperthermia**
  body temperature that has been raised abnormally high to kill tumour cells or make them more sensitive to the effects of radiation and certain chemotherapy drugs.

- **hyperthermic perfusion**
  a warmed solution containing chemotherapy drugs that bathes the tumour. It is used to try to shrink tumours and relieve symptoms.

- **hyperuricemia**
  a build up in the blood of uric acid (a byproduct of metabolism), a side effect of some chemotherapy drugs.

- **hyperuricosuria**
  abnormally high urinary uric acid.

- **hypervitaminosis d**
  an uncommon condition associated with kidney stones. caused by excessive or inappropriate Vitamin D intake.

- **hypocalcemia**
  see **parturient paresis.**

- **hypoglycemia**
  a less than normal amount of glucose in the blood, usually caused by administration of too much insulin, excessive secretion of insulin by the islet cells of the pancreas, or dietary deficiency. The condition may result in weakness, headache, hunger, visual disturbance, ataxia, anxiety, personality changes, and if untreated, delirium, coma and death.

- **hypoglycemia**
  below normal blood glucose level.

- **ICA**
  International Cooperative Alliance.

- **ICE**
  the abbreviation for the combination chemotherapy drugs ifosfamide, cyclophosphamide and etoposide.
- **ice milk**
  a frozen product resembling ice cream, except that it contains less fat (2 to 5 percent versus 10 percent) and more SNF (12 percent versus 10 percent) than ice cream. Both ice milk and ice cream contain stabilisers and emulsifiers and about 15 percent sugar.
- **ici d 1 694**
  an investigational chemotherapy drug that inhibits tumour cells from multiplying by interfering with cells' ability to make DNA. Also called raltitrexed and Tomudex.
- **idarubicin**
  a drug used in cancer chemotherapy. Also known as Idamycin.
- **idiopathic**
  from an unknown cause.
- **idoxuridine**
  a drug that prevents the growth of cancer cells by interfering with the cells' DNA.
- **ileostomy**
  the result of intestinal surgery where the far end of the, mall intestine is brought out to the skin surface. The small bowel contents are then collected in a bag attached to the skin. This is normally done only for severe large bowel disease because the fluid, bicarbonate and electrolyte losses can be difficult to treat or correct, especially on a long term basis.
- **imagery**
  a technique where the patient focuses on positive images to try to help the body to fight cancer and to feel better.
- **images source/ midwest plan service**
  herringbone parlour.
- **imaging procedures**
  methods of producing pictures of areas inside the body.
- **imitation milks**
  mixtures of nondairy ingredients (other than milk, milk fat, and nonfat milk solids) which are combined forming a product similar to milk, low fat milk, or skim milk. Sodium

caseinate, though derived from milk, is commonly termed a nondairy ingredient and is often used as a source of protein in imitation milks. Vegetable oils are commonly used as the source of fat.

- **immune function**
production of cells that fight disease or infection.

- **immune response**
the response of the immune system to challenge by a foreign antigen.

- **immune serum**
blood serum containing antibodies.

- **immune system**
the combination of cells, biological substances (such as antibodies) and cellular activities that work together to provide resistance to disease.

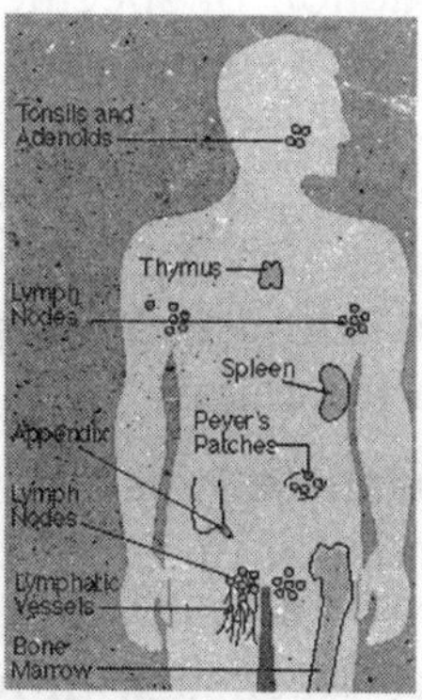

- **immunity**
1. nonsusceptibility to a disease or to the toxic effects of antigenic material. See **Active immunity**, **Cell-mediated immunity**, **Natural passive immunity**.
2. the power an animal has to resist and/or overcome an infection to which most of its species are susceptible. Active immunity is due to the presence of antibodies formed by an animal in response to previous exposure to the disease or through live or modified-live vaccines. Passive immunity is produced by giving the animal preformed or synthetic antibodies as with killed vaccines.

- **immunoassay**
technique for identifying substances based on the use of antibodies.

- **immunocompromised**
having a weakened immune system caused by certain diseases or drugs. This condition is sometimes present in the very young and the

very old, in patients with HIV, and in patients who have received a bone marrow transplant.

- **immunodeficiency syndrome**
the inability of the body to produce an immune response.

- **immunodiagnostics**
the use of specific antibodies to measure a substance. This tool is useful in diagnosing infectious diseases and the presence of foreign substances in a variety of human and animal fluids (blood, urine, etc.). It is currently being investigated as a way of locating tumour cells in the body.

- **immunofluorescence**
technique for identifying antigenic material that uses an antibody labeled with fluorescent material. Specific binding of the antibody and antigen can be seen under a microscope by applying ultraviolet light rays and noting the visible light that is produced.

- **immunogen**
any substance that can elicit an immune response.

- **immunoglobulin**
general name for proteins that function as antibodies. These proteins differ somewhat in structure and are grouped into five categories on the basis of these differences; immunoglobulin G (IgG), IgM, IgA, IgE and IgD.

- **immunologic adjuvant**
a drug that stimulates the ability of the immune system to respond to disease.

- **immunology**
study of all phenomena related to the body's response to antigenic challenge (i.e., immunity, sensitivity and allergy).

- **immunomodulators**
a diverse class of proteins that boost the immune system. Many are cell growth factors that accelerate the production of specific cells that are important in mounting an immune response in the body. These proteins are being investigated for use in possible cures for cancer.

- **immunoscintigraphy**
an imaging procedure involving the infusion of antibodies labelled with radio-

active substances followed by an imaging scan.

- **immunosuppression**
prevention or inhibition of the immune system to respond to foreign substances in the body.

- **immunosuppressive therapy**
therapy used to decrease the body's immune response, such as drugs given to prevent transplant rejection.

- **immunotherapy**
treatment to stimulate or restore the ability of the patient's immune system to fight infection and destroy tumour cells. Also called biological therapy.

- **immunotoxins**
specific monoclonal antibodies that have a protein toxin molecule attached. The monoclonal antibody is targeted against a tumour cell, and the toxin is designed to kill that cell when the antibody binds to it.

- **implantable pump**
a small device installed under the skin to administer a steady dosage of drugs.

- **impoundment**
confine in a reservoir.

- **improved grassland**
agriculturally improved grassland is grassland whose yield of grass is increased by human intervention. This may take the form of the application of fertilisers and/or herbicides and/or ploughing the land and re-seeding it with faster-growing grass varieties. In the case of wet areas, the land may be drained. While such measures will increase grassland productivity it usually results in a decrease in the plant species-richness.

- **in situ**
on site. In its original or natural place or position.

- **in situ cancer**
early cancer that has not spread to any other tissue.

- **in vitro**
1. literally, 'in glass'. Performed in a test tube or other laboratory apparatus.

2. in the laboratory (outside the body). The opposite of in vivo (in the body)
3. within an artificial environment, as within a test tube.

- **in vivo**
in the living organism.

- **inbreeding**
1. when two closely-related animals breed.
2. breeding between closely-related parents.

- **in-bye**
enclosed grassland close to a farm.

- **in-calf**
term used when a cow is pregnant.

- **incomplete record**
a cow's production in a lactation that is terminated early for some reason. (Lactations not yet completed may be referred to as a 'record in progress.')

- **incubated**
grown in the laboratory under controlled conditions. For instance, white blood cells can be grown in special conditions so that they attack specific cancer cells when returned to the body.

- **individual handler pool**
see **pool**.

- **indolent lymphoma**
lymphoma that is slow growing and has few symptoms.

- **induced lactation**
a lactation that is artificially initiated by the use of hormones, not by birth. In milk Designating a lactating female.

- **inducer**
a molecule or substance that increases the rate of enzyme synthesis, usually by blocking the action of the corresponding repressor.

- **induction therapy**
treatment designed to be used as a first step toward shrinking the cancer and in evaluating response to drugs and other agents. Induction therapy is followed by additional therapy to eliminate whatever cancer remains.

- **inedible**
adulterated, uninspected or not intended for use as human food.

- **inflammation**
swelling caused by the accumulation of lymph and blood cells at the site of infection or injury.

- **inflammatory bowel disease**
a general term that refers to inflammation of the colon and rectum found in ulcerative colitis and Crohn's disease.

- **infusion**
the introduction of a fluid, including medications, into the blood stream. Also called intravenous infusion.

- **ingredient sourcing**
finding the right ingredient(s) when producing a benchtop formulation.

- **inorganic**
not capable of sustaining life. Often refers to dirt or soil.

- **insect**
the largest class of arthropods. An insect's body is divided into three parts: head, thorax and abdomen. Three pairs of jointed legs are borne on the thorax, which is the location in many insects for a pair of wings.

- **insecticide**
a chemical substance that kills insects.

- **insectivorous**
insect-eating.

- **insemination**
the process of putting semen into a cows uterus to cause pregnancy.

- **intake protein (IP)**
total protein or nitrogen consumed without regard to quality.

- **intensification**
see **agricultural intensification.**

- **intensive agriculture**
see **agricultural intensification.**

- **interferon**
1. a class of lymphokine proteins important in the immune response. There

are three major types of interferon: alpha (leukocyte), beta (fibroblast) and gamma (immune). Interferons inhibit viral infections and may have anticancer properties.

2. interferons are a type of biological response modifier (a substance that can improve the body's natural response to disease). Interferons interfere with the division of cancer cells and thus slow the growth of the tumour. There are several types of interferon, including interferon alfa, beta and gamma.

- **interleukin**
  1. a type of lymphokine that regulates the growth and development of white blood cells. Twelve interleukins (IL-1 through IL-12) have been identified to date.
  2. a substance used in biological therapy to help the immune system fight infection and cancer.

- **interleukin-11**
  a substance that stimulates immune response and may reduce toxicity to the gastrointestinal system resulting from cancer therapy. Also called IL- I 1.

- **interleukin-12**
  a substance that enhances the ability of the immune system to kill tumour cells, and that may interfere with blood flow to the tumour. Also called IL- 1 2.

- **interleukin-2**
  a type of biological response modifier (a substance that can improve the body's natural response to disease) that stimulates the growth of certain disease-fighting blood cells in the immune system. Also called IL-2.

- **interleukin-4**
  a substance that enhances the immune system's ability to fight tumour cells. Also called IL-4.

- **intermediate-grade lymphomas**
  includes diffuse small, cleaved cell lymphoma and diffuse large, noncleaved cell lymphoma. These are more aggressive than low-grade lymphomas, but they have a high response rate to combination chemotherapy.

■ **internal temperature**
the temperature of the internal portion of a food product.

■ **interpretation**
provision of information about a site or area (e.g. through fixed signs).

■ **intracarotid**
the carotid artery is the main artery in the neck; it carries blood from the heart to the brain. An intracarotid infusion is the introduction of fluids and drugs directly into the carotid artery.

■ **intramuscular**
an injection into the muscle. Usually abbreviated IM.

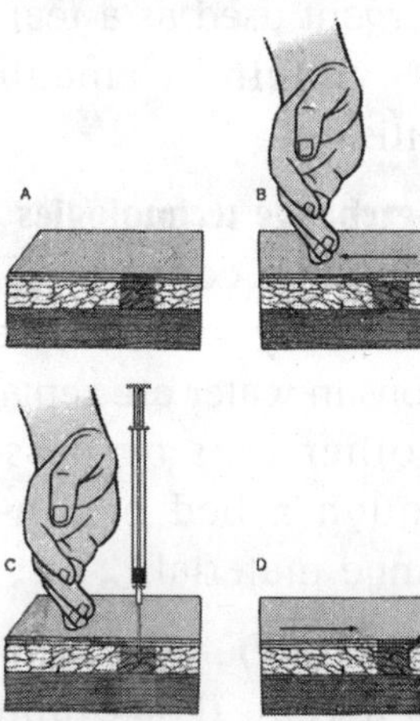

■ **intrathecal**
the thin space between the lining of the spinal cord and brain. Chemotherapy drugs can be infused into the spinal fluid there to treat or prevent cancers in the brain and spinal cord.

■ **intravenous pyelogram (IVP)**
this is a series of X-rays designed to show the inside of the kidneys and urinary system. Traditionally, this was the standard diagnostic tool for evaluation of possible kidney stones and tumours. A small amount of contrast or dye is injected into a vein. This contrast is excreted by the kidneys within five to fifteen minutes. When the contrast appears in the urine, it clearly outlines the shape, size and location of the internal urinary organs. If there is blockage or dilation, it will show up clearly. Virtually all significant stones will be visible either directly on the X-ray or indirectly through the blockage they produce.

■ **intraventricular**
the delivery of a drug into a space within an organ.

■ **intron**
in eukaryotic cells, a sequence of DNA that is con-

tained in the gene but does not encode for protein. The presence of introns 'splits' the coding region of the gene into segments called exons. See **Exon**.

- **invasive**
  colonising an area from outside.

- **invasive cancer**
  cancer that has spread beyond its site of origin and is growing into surrounding, healthy tissues. Also called infiltrating cancer.

- **invert sugar**
  means a mixture of glucose (dextrose) and fructose (levulose) formed by the hydrolysis of sucrose.

- **invertebrate**
  an animal without a backbone.

- **investigational new drug application (IND)**
  an application to begin studies of a new drug or biologic on humans. The IND gives the plan for the study and contains formulation, manufacturing and animal test result information.

- **involution**
  a decline in size or activity of tissues and/or organs. For example, the mammary gland tissues normally involute with advancing lactation. It is part of the process of lactating cows wherein they dry off.

- **iodinated casein**
  milk protein (casein) to which iodine is bound. It is commonly referred to a thyroprotein and can be used to stimulate dairy cows to secrete more milk.

- **iodophor**
  combination of iodine and detergent used as a teat dip or milking machine sanitiser.

- **ion-exchange technologies**
  technology concerning the process by which certain anions in water are replaced by other ions by passage through a bed of ion-exchange material.

- **iravenous (IV)**
  into a vein. (For example, an intravenous medication would be injected into a vein).

■ **irinotecan**
belongs to a group of anti-cancer drugs known as camptothecin analogues. Also knows as CPT-1 1.

■ **irradiation**
the process where foods, such as poultry, red meat, spices, and fruits and vegetables, are subjected to small amounts of radiant energy including gamma rays, electron beams and x-rays.

■ **irritable bowel syndrome**
abnormally increased motility of the small and large intestines generally associated with emotional stress. Most of those affected are young adults, who complain of diarrhoea and, occasionally, pain in the lower abdomen. The pain is usually relieved by moving the bowels. In diagnosing irritable bowel syndrome other more serious conditions, such as dysentery, lactose intolerance, and the inflammatory bowel disease, must be ruled out. Because there is no organic disease present in IBS, no specific treatment is necessary. Many persons benefit from the use of bulk-producing agents in the diet, because bulk tends to stabilise the water content of the stool. Antidiarrheal drugs are helpful in decreasing the frequency of the stool. Also called functional bowel syndrome, mucous colitis and spastic colon.

■ **isis 3521**
an anticancer drug that inhibits tumour growth.

■ **isis 5132**
an anticancer drug that inhibits tumour growth.

■ **islet cells**
pancreatic cells that are the source of insulin and two other hormones involved in regulating glucose metabolism and absorption.

■ **ISO accreditation**
accreditation by the International Organisation for Standardisation (ISO), which develops international agreements on standards for various industries.

■ **isoenzyme**
one of the several forms that a given enzyme can take. The forms may differ in certain physical properties, but

function similarly as biocatalysts.

- **isogenic**
  of the same genotype.
- **isosulfan blue**
  a dye used in identifying sites of tumour development.
- **IU (international unit)**
  a unit of measurement of a biologic (e.g., a vitamin, hormone, antibiotic, antitoxin) as defined by the International Conference for Unification of Formulae. Potency is based on bioassay that produces a particular effect agreed on internationally. Also called USP unit.
- **jerky**
  this product is a nutrient-dense meat that has been made lightweight by drying. Products may be cured or uncured, dried, and may be smoked or unsmoked, air or oven dried.
- **kamenitza**
  a small depression, a few m in diameter and several cm deep, in a level surface of calcareous rock like limestone. These features are caused by standing water slowly dissolving the rock.
- **kaposi's sarcoma**
  a relatively rare type of cancer characterised by the abnormal growth of blood vessels that develop into skin lesions or occur internally. It is common for people with AIDS to have Kaposi's sarcoma.
- **kappa-casein**
  (molecular weight 19,000; 169 residues, 20 prolines) very resistant to calcium precipitation, stabilising other caseins. Rennet cleavage at the Phe105-Met106 bond eliminates the stabilising ability, leaving a hydrophobic portion, para-kappa-casein, and a hydrophilic portion called kappa-casein glycomacropeptide (GMP), or more accurately, caseinomacropeptide (CMP).
- **keratinocyte growth factor**
  stimulates the growth of epithelial cells that line the surface of the mouth and intestinal tract.
- **ketonuria**
  the presence of ketone bodies in the urine, as with ketosis in high-producing cows.

- **keyhole limpet hemocyanin**
  one of a group of drugs known as immune modulators, given as a vaccine to help the body respond to cancer. Also called KLH.

- **kid**
  a young goat. A goat up to one year old.

- **kidding parturition**
  in the goat, the act of giving birth to young.

- **kidney plasminogen activator**
  a precursor to the enzyme urokinase that has blood-clotting properties.

- **killer cells**
  blood cells that attack tumour cells and body cells that have been invaded by foreign substances.

- **kilocalorie (kcal)**
  equivalent to 1000 calories.

- **kissle**
  one of the newest products in the dairy case. It is a creamy blend of milk and fruit, in a yogurt-type container, with a sweeter taste than most yogurt. Kissle is made from many of the same ingredients as fruit-based yogurt but does riot contain yogurt culture, preservatives, or artificial flavor. Kissle can be refrigerated up to 90 days.

- **kub**
  stands for kidneys, ureters and bladder. This is just medical jargon for a flat X-ray of the abdomen. Calcium stones often show up well on this type of X-ray.

- **label**
  a display of written, printed, or graphic matter upon the immediate container (not including package liners) of any food product.

- **lachrymation**
  the act of tearing; secreting and conveying tears.

- **lactaid**
  one brand of lactose-reduced milk. It is a 1% low- fat milk product that has had 70% of its lactoge broken down. It appeals to people who cannot digest milk easily. LactAid is UHT pasteurised for a long shelf life.

- **lactalbumin**
  a whey protein. A soluble milk protein concentrate.

- **lactate**
  1. to secrete or produce milk.

2. to secrete or produce milk; also, salt of beta-hydroxy propionic acid.

- **lactation**
  the process of secreting milk.
- **lactation average**
  actual 305-day average milk and component production for all cows in the herd that was completed during a 12-month period. (Since dry days are not included, lactation averages usually are 8 to 15 percent higher than rolling herd averages.)
- **lactation period**
  the number of days an animal secretes milk following each parturition. Usually 11 to 12 months.
- **lactation record**
  the total milk and components produced by a cow beginning on the day of calving and ending on the day the cow goes dry. For purposes of genetic comparison, 10-month (305-day) lactation records are the standard of the industry. A 306 to 365-day lactation record may be used for promotion. Lactation records greater than 365 days will not be published.
- **lactation totals to date**
  the production totals for milk and components through the current test date.
- **lactogenic**
  stimulating the secretion of milk.
- **lactoglobulin**
  a whey protein.
- **lacto-ovo-vegetarian**
  one whose diet consists primarily of foods that are vegetable origin and also includes some animal products, such as eggs (ovo), milk, and cheese (lacto), but no meat, fish, or poultry.
- **lactoperoxidase**
  enzyme that keeps the milk for a longer period.
- **lactose**
  1. type of sugar, a disaccharide, naturally found in milk.
  2. white, odourless crystalline sugar present in milk.

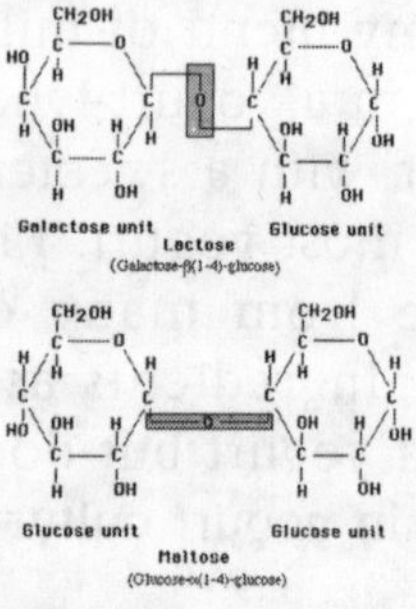

3. a disaccharide, $C_{12}H_{22}O_{11}$, found in milk, which may be hydrolysed to yield glucose and galactose. A white crystalline substance obtained from whey and used in infant foods, bakery products, confections and pharmaceuticals as a diluent and excipient. Also called milk sugar. See **dairy**, **lactose intolerance**.
4. lactose is a disaccharide (2 sugars) made up of glucose and galactose (which are both monosaccharides). It comprises 4.8 to 5.2% of milk, 52% of milk SNF, and 70% of whey solids. It is not as sweet as sucrose. When lactose is hydrolysed by ß -D-galactosidase (lactase), an enzyme that splits these monosaccharides, the result is increased sweetness and depressed freezing point.

One of its most important functions is its utilisation as a fermentation substrate. Lactic acid bacteria produce lactic acid from lactose, which is the beginning of many fermented dairy products. Because of their ability to metabolise lactose, they have a competitive advantage over many pathogenic and spoilage organisms. Some people suffer from lactose intolerance; they lack the lactase enzyme, hence they cannot digest lactose, or dairy products containing lactose. Crystallisation of lactose occurs in an alpha form which commonly takes a tomahawk shape. This results in the defect called sandiness. Lactose is relatively insoluble which is a problem in many dairy products, ice cream, sweetened condensed milk. In addition to lactose, fresh milk contains other carbohydrates in small amounts, including glucose, galactose, and oligosaccharides.

5. the 12-carbon sugar unique to milk; a disaccharide composed of glucose and galactose. Average milk contains about 5 percent lactose which varies little by breed.

lactose 'milk sugar'
a chemical found naturally in milk. Lactose-reduced products am available for people who have trouble digesting lactose. Lactose in milk can be reduced by 70% when lactose—a chemical available in liquid or tablet form — is added to milk and refrigerated for 24 hours. Other options for consumers include products such as Easy 2% (2% low-fat milk) and LactAid (1% low-fat milk), which already have 70% of the lactose broken down.

lactose intolerance
1. a sensitivity disorder resulting in the inability to digest lactose because of a deficiency of or defect in the enzyme lactase. Symptoms of the disorder are bloating, flatus, nausea, diarrhoea and abdominal cramps. The diet is adjusted according to the tolerance level, restricting such milk-derived foods as milk, cheese, butter, margarine, and any products containing milk, such as cakes, ice cream, cream soups and sauces. See **Dairy, Lactose.**
2. the inability to digest or absorb lactose, a type of sugar found in milk and other dairy products.
3. a condition in which lactose is not adequately hydrolysed in the intestine because of insufficient lactase enzyme.

lagoon
an earthen pond used as a primary storage site for manure.

lamb
1. a term that refers to both young sheep and the meat obtained from sheep that are generally slaughtered within 12-14 months of birth.
2. meat from sheep less than one year old. If the phrase 'Spring Lamb' is on a meat label, it means the lamb was produced between March and October.

■ **laparoscopy**
the insertion of a thin lighted instrument (a laparoscope) through the abdominal wall to inspect the inside of the abdomen and perform biopsies.

■ **laparotomy**
a surgical incision made into the wall of the abdomen.

■ **lard**
lard is the fat rendered from clean and sound edible tissues from swine.

■ **larva**
a stage in the life cycle of an insect following emergence from the egg in those species where the young insect is markedly different from the adult. A good example is a caterpillar.

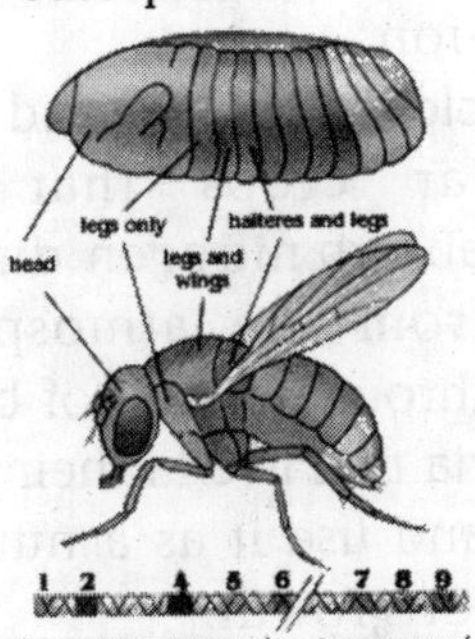

■ **laser**

1. most people now know what a laser is: a powerful, amplified beam of light that can cut through steel or tissue with ease. The laser discussed in this book is specially tuned to affect only materials that form stones. Normal tissue is not harmed. The laser is a powerful tool to fragment stones, but it requires direct contact with the stone through some sort of telescope in either the kidney, ureter or bladder.
2. an intensely powerful beam of light that is used in surgery, to diagnose disease and in cancer therapy.

■ **laser therapy**
the use of a narrow beam of light to kill cancer cells.

■ **leaching**
the process of removal of soluble materials by the passage of water through soil.

■ **lead feeding**
gradually increasing grain fed cows to a level equal 1.0 to 1.5 percent of body

weight beginning about three weeks prior to the predicted calving date. Following parturition, grain is increased until a cow reaches maximum feed intake. Also called challenge feeding.

- **leaf litter**

an accumulation of dead plant remains on the soil surface.

- **leaf margin**

the edge of a leaf.

- **leaky gut syndrome**

a term used to describe a condition in which the lining of the intestinal tract becomes perforated and irritated, and tiny particles of partially digested food enter the bloodstream, causing an allergic reaction.

- **lecithin**

a fatty substance that is present in plant and animal tissue.

- **leflunomide**

an anticancer drug.

- **leg bands**

cloth or plastic strips of a bright colour used for marking treated cows, fresh cows or cows needing special handling.

- **legume**

1. any of thousands of plant species that have seed pods that split along both sides when ripe. Some of the more common legumes used for human consumption are beans, lentils, peanuts, peas, and soybeans. Others, such as clover and alfalfa, are used as animal feed. Legumes have a unique ability to obtain much or all of their nitrogen requirements from symbiotic nitrogen fixation.
2. clovers, alfalfa and similar crops that can absorb nitrogen directly from the atmosphere through action of bacteria that live in their roots and use it as a nutrient for growth.

- **leptomeningeal cancer**
  involves the tissues that cover the brain and spinal cord.

- **leptomeningeal metastases**
  tumour cells that have spread from the original (primary) tumour to the tissues that cover the brain and spinal cord.

- **less favoured area (LFA)**
  areas (usually in the uplands) where agricultural subsidies are available to livestock farmers to help maintain economically-viable farming in parts of the countryside with low returns. The LFAs are split into disadvantaged areas (DAs) and severely disadvantaged areas (SDAs) with corresponding rates of payment (higher on the whole in SDAs than in DAs.

- **letdown**
  the process in a cow where physical stimulation causes a release of oxytocin and the contraction of smooth muscles surrounding milk alveoli resulting in fluid pressure within the udder and milk flow.

- **leucovorin**
  a drug used to protect normal cells from high doses of the anticancer drug methotrexate. It is also used to increase the antitumour effects of the drugs fluorouracil and tegafur-uracil, an oral treatment alternative to intravenous fluorouracil.

- **leukapheresis**
  removal of the blood to harvest specific blood cells, the remaining blood is returned to the body.

- **leukemia**
  cancer of the blood and bone marrow.

- **leukocyte**
  a colorless cell in the blood, lymph and tissues that is an important component of the body's immune system. Also called white blood cells.

- **leukoplakia**
  a precancerous lesion that may develop on mucous membranes.

- **levamisole**
  a drug that restores immune function.

- **leveret**
  a young hare.

levodopa
an investigational drug used in cancer chemotherapy.

ley
a short term grass pasture, often cut for silage or hay. It is sown to last for one or more years, after which it is ploughed up and replaced with another crop.

libido
sexual desire or instinct.

library
a set of cloned DNA fragments that taken collectively contain the entire genome of an organism. Also called a DNA library.

lichen
a composite plant comprised of an alga and a fungus growing together in an intimate association. Lichens are often found in patches on tree trunks, bare ground, rocks and walls. They can be sensitive indicators of pollution.

life cycle
the various stages of development through which organisms of a species pass from the fertilised egg of one generation through to the same stage in the next generation.

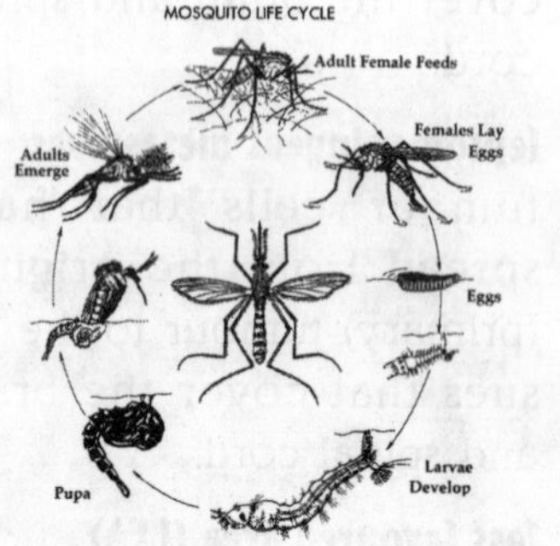

life expectancy
the length of time that a particular organism is expected to live.

ligase
an enzyme used to join DNA or RNA segments together. They are called DNA ligase or RNA ligase, respectively.

light cheese
product that is manufactured to have different sodium and/or fat content.

light cream (coffee cream, table cream)
cream that contains between 18% and 30% milk fat and may be homogenised.

lignin
a compound which, with cellulose, forms the cell walls of plants. It is practically indigestible.

▪ **linear score (SCC)**
linear scores for somatic cell counts (SCC's) convert SCC logarithmically from cells per mm to a linear score from 0 to 9. The linear score has a straight line, inverse relationship with milk yield. An increase of one in the linear score is associated with a 400-pound decrease in lactation milk yield or a 1.5 pound drop in daily yield.

▪ **liner**
a flexible sleeve in the milking teat cup or rigid-walled liner holder. Responsible for massaging the teat end and intermittently cutting vacuum at the teat end during milking. Also called inflation.

▪ **liner slip**
slippage of the liner and teat cup during milking. Caused a sharp change of milking vacuum within the unit or cluster.

▪ **liner slips or squawks**
slippage of the liner and teat cup during milking. Caused by a sharp change of milking vacuum within the unit or cluster by drawing in air alongside the teat. Generally creates a 'squawking' sound.

▪ **linkage**
the tendency for certain genes to be inherited together due to their physical proximity on the chromosome.

▪ **linker**
a fragment of DNA with a restriction site that can be used to join DNA strands.

▪ **lipid**
1. the main type of fat found in the body.
2. any one of a group of organic substances that are insoluble in water though soluble in alcohol, ether, chloroform, and other fat solvents, and have a greasy feel. They are rich sources of dietary energy.

▪ **lipoprotein lipase (LPL)**
a lipase enzyme splits fats into glycerol and free fatty acids. This enzyme is found mainly in the plasma in association with casein micelles. The milk fat is protected from its action by the FGM. If the FGM has been damaged, or if certain cofactors (blood serum lipopro-

teins) are present, the LPL is able to attack the lipoproteins of the FGM. Lipolysis may be caused in this way.

- **lipoproteins**
  a class of serum proteins that transport lipids and cholesterol in the blood stream. Abnormalities in lipoprotein metabolism have been implicated in certain heart diseases.

- **liposomal**
  a drug preparation that contains the active drug in very tiny fat particles. This fat-encapsulated drug is absorbed well and its distribution to the tumour site is improved.

- **liquid nitrogen**
  super cold (-196 degrees C) liquefied gas.

- **liquid sugar**
  means a direct consumption sugar which is not principally of crystalline structure and which contains, or which is to be used for the production of, any sugars principally not of crystalline structure which contain soluble non-sugar solids (excluding any foreign substances that may have been added or developed in the product) equal to 6 percent or less of the total soluble solids. Liquid sugar is exclusive of cane syrup and edible molasses.

- **liquid-solid separation**
  the process of passing a liquid-solid suspension, such as animal manure, over a screen or similar device to partially remove solids prior to storage or application.

- **lisofylline**
  an investigational drug that may protect healthy cells from chemotherapy and radiation without inhibiting the effects of these therapies on tumour cells.

- **listeria**
  1. a bacteria that causes an infectious disease transmitted through contaminated food, including raw meat, poultry and milk.
  2. listeria monocytogenes, a pathogenic bacterium that can be carried in a variety of foods such as dairy products, red meat, poultry, seafood and vegetables.

- **lithoclast**
  the proper name of a particular kidney stone fragmenting probe that uses a small version of a pneumatic hammer to fragment stones. Works very well and quite safe but relatively new and only now becoming widely available. More popular in Europe.

- **lithotripsy**
  the name for any procedure, surgery or technique that fragments or breaks up a stone. Often refers to ESWL.

- **lithotriptor**
  literally meaning 'stone breaker', it refers to any machine than can fragment a stone; usually refers to some type of ESWL machine.

- **liverwort**
  a group of bryophytes, the members of which are not differentiated into stem and leaves (Compare with mosses).

- **livestock**
  1. animals kept on a farm (e.g. sheep and cows).
  2. cattle, sheep, swine, goat, horse, mule or other equine.

- **livestock unit**
  method of describing different livestock types and age groups based on their energy requirements. Standard ratios are used, commonly based around one livestock unit equalling one Friesian dairy cow.

- **liveweight**
  the weight of an animal.

- **loafing (area) (alley) (barn) (shed)**
  a resting area for cattle.

- **local agenda 21 (LA21)**
  at the earth summit a programme of local action was agreed to achieve Sustainable Development in the twenty-first century. This was 'Agenda 21'. LA21 is the UK response to this with local government expected to take the lead in promoting action at a local community level.

- **local therapy**
  treatment that affects cells in the tumour and close to it.

- **localised**
  restricted to the site of origin without evidence of spread.

- **locally advanced cancer**
  cancer that has spread only to nearby tissues or lymph nodes.

- **log reduction**
  'log' stands for logarithm, which is the exponent of 10. For example, log2 represents 10^2 or 10 x 10 or 100. Log reduction stands for a 10-fold or one decimal or 90% reduction in numbers of recoverable bacteria in a test food vehicle. Another way to look at it is: 1 log reduction would reduce the number of bacteria 90%. This means, for example, that 100 bacteria would be reduced to 10 or 10 reduced to 1.

- **long feed**
  coarse or unchopped feed such as hay, as contrasted with short or chopped, feed.

- **loose housing**
  facilities that allow cattle access to a large, open bedded area for resting (also known as free housing). Loose housing should provide at least 200 square feet per animal for feeding and resting (freestall housing uses only 90 square feet per animal). See **Open Barns**.

- **loperamide hydrochloride**
  an antidiarrheal drug. Also called Imodium.

- **losoxantrone**
  belongs to a group of chemotherapy drugs known as antipyrazoles.

- **low risk-near threatened**
  a relatively low risk category that describes the status of a particular SPECIES. Compare with endangered, vulnerable or rare.

- **low fat milk**
  1. milk containing at least 8.25 percent solids and from which sufficient milk fat has been removed to produce, within limits of good manufacturing practice, a milk fat content of 0.5, 1.0, or 2.0 percent.
  2. milk labeled low-fat, 2% milk and 1% milk. Lowfat milk contains between 0.5% and 2.0% milkfat and at least 8.25% nonfat solids. Low-fat milk, which contains at least 10% nonfat solids most, is labeled 'protein fortified'

or 'fortified with protein'. Vitamin A is also added to low-fat milk to replace the Vitamin A, which is lost during the manufacturing process.

- **low-grade lymphomas**
  includes chronic lymphocytic lymphoma and follicular small cleaved cell lymphoma. Patients with no symptoms, even with widespread disease, may require no initial treatment. These disorders are associated with a high response rate to treatment.

- **low-intensity farming**
  farming practices that do not employ many of the intensive methods. See **agricultural intensification**.

- **low-moisture silage**
  silage that conta ins 35 to 55 percent moisture.

- **lu-103793**
  an investigational chemotherapy drug that prevents the growth and reproduction of tumour cells.

- **lucerne alfalfa (medicago sativa)**
  a legume of high feeding value for ruminants.

- **lumpectomy**
  removal of the tumour and a small amount of normal tissue around it.

- **lymph node drainage**
  the area of tissue that drains lymphatic fluid into a particular lymph node.

- **lymph node mapping**
  the use of dyes and radioactive substances to identify lymph nodes that contain tumour cells.

- **lymph nodes**
  small, bean-shaped organs located throughout the body along the channels of the lymphatic system. Bacteria or cancer cells that enter the lymphatic system may be found in the lymph nodes (also called lymph glands). Lymph nodes associated with the organ in which the cancer is found are often examined to see if the cancer has spread.

- **lymphadenectomy**
  lymphadenectomy (also known as lymph node dissection) is a surgical procedure in which the lymph nodes are removed and examined to see if they contain cancer.

■ **lymphatic system**
the tissues and organs that produce and store cells that fight infection and disease.

■ **lymphocyte**
1. a type of leukocyte found in lymphatic tissue in the blood, lymph nodes and organs. Lymphocytes are continuously made in the bone marrow and mature into antibody-forming cells. See **B-lymphocytes**.
2. white blood cell. Lymphocytes have a number of roles in the immune system, including antibody production, attacking and destroying cancer cells, and producing substances that kill cancer cells.

■ **lymphography**
an x-ray of lymph nodes and lymphatic vessels after injection of a dye.

■ **lymphokine**
a class of soluble proteins produced by white blood cells that play a role, as yet not fully understood, in the immune response. See **Interferon, Interleukin**.

■ **lymphokine-activated killer cells**
white blood cells that are stimulated in a laboratory to kill tumour cells. Also called LAK cells.

■ **lymphoma**
1. form of cancer that affects the lymph tissue.
2. cancer of cells of the lymphatic system, a network of lymph vessels, lymph nodes, tonsils and the spleen, which are involved in fighting infection.

■ **lymphomatoid granulomatosis**
destructive growth of lymph cells, usually involving the lungs, skin, kidneys and central nervous system. Grades I and II are not considered cancerous, but grade III is considered a lymphoma.

■ **lymphoproliferative disorders**
diseases of the lymphatic system.

■ **macrophage**
a type of white blood cell produced in blood vessels and loose connective tissues that can ingest dead tissues and cells and is involved in producing interleukin-1. When exposed to the lymphokine macrophage-acti-

vating factor, macrophages also kill tumour cells.

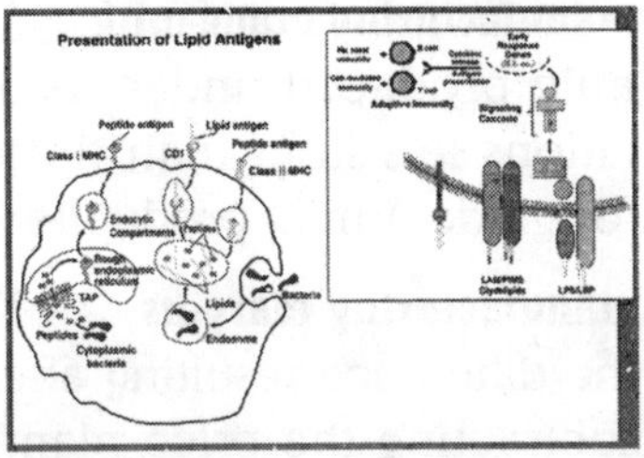

■ **macrophage colony stimulating factor (m-csf)**
a natural hormone that stimulates the production of white blood cells, particularly monocytes (the precursors of macrophages).

■ **macrophyte**
an aquatic plant that can be seen without the aid of a microscope.

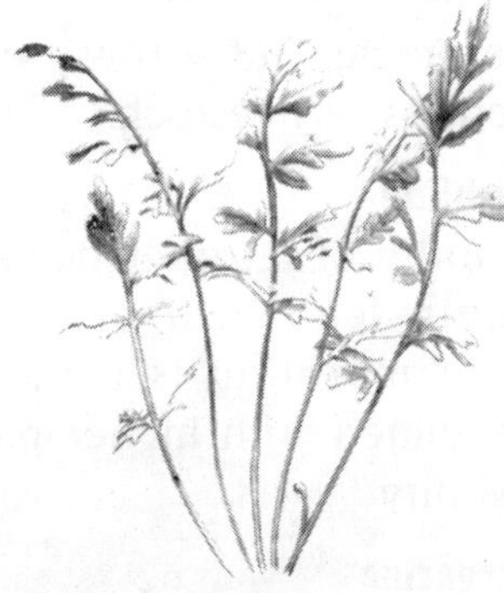

■ **mad cow disease**
the common term used for Bovine Spongiform Encephalopathy (BSE).

■ **maillard reaction**
lactose and proteins in milk react with other reducing sugars during baking or cooking to produce a desirable browning colouration and caramel flavouring.

■ **maintenance therapy**
treatment that is given to help prevent relapse in patients whose cancer is in remission.

■ **malabsorption syndrome**
a group of symptoms resulting from the body's inability to properly absorb nutrients.

■ **malignancy**
a cancerous tumour that spreads to other parts of the body.

■ **malignant**
a malignant tumour is a cancerous growth with a tendency to spread to other parts of the body.

■ **malignant ascites**
a condition in which fluid collects in the abdomen (ascites) and contains cancer (malignant) cells.

■ **malt lymphoma**
mucosa-associated lymphoid tissue (MALT) lym-

phoma is a tumour of the cells involved in antibody production which are found in the mucosal tissue.

- **manufactured milk products**
these include cheeses, butter, evaporated whole milk, condensed whole and skim milks, whole milk powder, nonfat dry milk, ice cream, frozen desserts, aerated cream, frozen and plastic creams and milk used in candy, soup, bakery products, and animal feeds.

- **manufactured products**
includes basically all dairy products except fluid milks: all cheeses, butter, evaporated whole milk, condensed whole milk, condensed skim milk, whole milk powder, non-fat dry milk, ice cream, ice cream mix, frozen desserts, creams and whey products.

- **manufacturing cooperative**
association of milk producers who own and operate milk manufacturing and processing facilities and market the dairy products for its members. A manufacturing cooperative typically also sell portions of its milk supply to other handlers.

- **manufacturing grade milk**
milk produced under regulations less strict than those for grade A milk production.

- **manufacturing margins**
the difference resulting after subtracting the price plants pay for milk from the calculated wholesale value of dairy products produced by milk manufacturing plants. This margin is influenced by competitive marketing conditions at both the wholesale and farm level markets.

- **manufacturing milk**
milk for the manufacture of dairy products.

- **manure**
the faecal and urinary excretions of livestock.

- **marbling**
quantity of fat interspersed within lean beef muscle. More marbling is usually associated with higher palatability.

- **margarine**
a high quality food product that is often used as a substitute for butter in cooking and for table use. It differs

from butter in that its primary ingredient is vegetable oil while butter's primary ingredient is fresh milk.

- **marimastat**
  an investigational drug that is being studied for its ability to reduce and prevent tumour growth.

- **marinade**
  a savoury acidic sauce in which a food is soaked to enrich its flavour or to tenderise it. Marinade consists of a cooking oil, an acid (vinegar, lemon juice, wine), and spices. As the food stands in the mixture, the acid and the oil impart the savoury flavours of the spices to the food. The acid also has a tenderising action.

- **marinate**
  to steep food in a marinade.

- **marker**
  a diagnostic indication that disease may develop.

- **marl**
  a form of clay rich in calcium carbonate.

- **marl lake**
  a freshwater lake with concentrations of dissolved calcium carbonate greater than 100 mg/litre.

- **marsh**
  a tract of low, ill-drained ground with patches of open water in which, in temperate regions abound reeds, rushes and sedges.

- **mastectomy**
  removal of the mammary glands.

- **mastitis**
  an inflammation of the mammary gland (or glands), usually caused by bacteria.

- **maternity barn**
  a special needs facility where cows can be closely monitored during the period immediately before and after they give birth.

- **mating**
  the act of bringing a sperm together with an egg to produce an embryo and ultimately a calf.

■ **mattress**
bedding material compacted to 3 to 4 inches and sandwiched in a heavyweight polypropolene or other fabric. Possible fillers include: long or chopped straw, poor quality hay, sawdust, shavings, rice hulls and, most commonly, shredded rubber.

■ **mdm**
mechanically deboned meat.

■ **meadow**
see **hay meadow**

■ **measurable disease**
a tumour that can be accurately measured in size. This information can be used to judge response to treatment.

■ **meat**
the flesh of animals used as food including the dressed flesh of cattle, swine, sheep, or goats and other edible animals, except fish, poultry and wild game animals.

■ **meat base**
a granular, paste-like product which is shelf-stable primarily because of its high salt content (30-40%).

■ **meat byproduct**
any part capable of use as human food, other than meat, which has been derived from one or more cattle, sheep, swine, or goats. This term, as applied to products of equines, shall have a meaning comparable to that provided in this paragraph with respect to cattle, sheep, swine and goats.

■ **mechanically separated (species)**
mechanically separated meat (i.e., beef, veal, pork, lamb).

■ **mechanically separated poultry**
mechanically separated poultry (i.e., chicken, turkey) is a paste-like and batter-like product produced by forcing the bones and attached edible tissue through a sieve or similar device to separate the bone from the edible poultry tissue. This product is intended for use in the formulation of other poultry products. During this process, it is possible for bones to be crushed or pulverised, resulting in a limited amount of bone particles. Because it may contain some bone par-

ticles, any product that has been produced using the mechanical separation process cannot, by definition, be labelled as 'poultry'. Instead, it must be labelled appropriately as 'mechanically separated chicken' (or turkey) on the product's label. Mechanically separated chicken and turkey are used in products such as chicken and turkey franks, bologna, nuggets and patties.

- **mechlorethamine**
  a topical ointment for treatment of infections of the skin.

- **medial supraclavicular lymph nodes**
  lymph nodes located above the collar bone and between the center of the body and a line drawn through the nipple to the shoulder.

- **medium**
  a substance containing nutrients needed for cell growth.

- **medium cheese**
  cheese which is cured up to six months, is mellow bodied, has a smooth texture, and tastes sharper than mild cheese.

- **medullary sponge kidney**
  a benign condition of the kidneys caused by dilation of the microscopic renal collecting tubules. It's diagnosed by kidney X-rays (Intravenous Pyelogram or IVP) and is described as a faint, white blush on the inside of the kidney visible only on Xray. It's associated with an increased risk of stone disease but isn't otherwise dangerous.

- **meetings**
  get together of group members.

- **megacalorie (mcal)**
  equivalent to 1000 kcal or 1,000,000 cal. A megacalorie is equivalent to a therm.

- **megestrol**
  a drug that belongs to the group of hormones known as progestins, used as hormone therapy to block estrogen and to suppress the effects of androgens. It is also used to stimulate the appetite in cancer patients. Also known as Megace.

- **meiosis**
  process of cell reproduction whereby the daughter cells have half the chromosome

number of the parent cells. Sex cells are formed by meiosis. Compare Mitosis.

- **meningeal**
refers to the tissue covering the brain (the meninges) and spinal cord. Leptomeningeal cancer involves two of the three tissue layers that make up the meninges.

- **meningeal metastases**
tumour cells that have spread from the original (primary) tumour to the tissue covering the brain and/ or spinal cord.

- **mercaptopurine**
belongs to the group of anticancer drugs known as purine analogues. Also known as 6-MP.

- **mesenchymal**
cells that develop into connective tissue, blood vessels, and lymphatic tissue.

- **mesna**
a drug that protects healthy cells from the toxic effects of ifosfainide and cyclophosphamide.

- **mesophilic bacteria**
bacteria that thrive under high humidity.

- **mesotrophic**
moderately nutrient-rich.

- **messenger RNA (mRNA)**
nucleic acid that carries instructions to a ribosome for the synthesis of a particular protein.

- **metabolic testing**
a comprehensive package of kidney stone prevention blood and urine chemistry tests.

- **metabolic weight**
the weight of an animal raised to threequarter power (W0.75).

- **metabolism**
1. the sum total of all of the chemical reactions involved in living organisms.
2. all biochemical activities carried out by an organism to maintain life.

- **metabolisable energy (ME)**
food-intake gross energy minus fecal energy, energy in the gaseous products of digestion (mostly methane), and urinary energy.

- **metamorphose**
to undergo metamorphosis.

- **metamorphosis**
a stage in the life cycle of certain animals during which

the larva rapidly transforms into an adult. An example of metamorphosis is the change from caterpillar to adult moth or butterfly.

- **metaplasia**
  the change of cells in adult tissue from normal to abnormal.

- **metastatic cancer**
  cancer that has spread from the place in which it started to other parts of the body.

- **methodology**
  a systematic way of conducting a specific survey or piece of research.

- **metritis**
  an inflammation of the uterus.

- **MF (milk fat)**
  the lipid or fatty acid components of milk.

- **microbial herbicides and pesticides**
  microorganisms that are toxic to specific plants or insects. Because of their narrow host range and limited toxicity, these microorganisms may be preferable to their chemical counterparts for certain pest control applications.

- **microbiology**
  study of living organisms that can be seen only under a microscope.

- **microclimate**
  the climate within a few m of the ground and in a relatively small area.

- **microinjection**
  the injection of DNA using a very fine needle into a cell.

- **microorganism**
  1. a form of life that can be seen only with a microscope, including bacteria, viruses, yeast and single-celled animals.
  2. any organism that can be seen only with the aid of a microscope. Also called microbe.
  3. a microscopic organism.

- **micturate**
  another word for urinate or void.

- **migrate**
  1. to go on migration.
  2. the periodic movement of animals from one region to another usually connected with seasonal changes. In many animals, migration is a response to lower tempera-

tures and decreased supply of food in winter. In such cases, it may be regarded as an alternative strategy to hibernation.

3. transfer of a component of a packaging material into the product contained, or loss of a component of the product into the packaging material.

- **mild cheese**

cheese which is cured for two to three months, has little flavour development, and a soft texture.

- **milk**

1. milk from animals that is collected and processed (from cows, goats, sheep, yaks, buffaloes, camels).
2. the lacteal secretion from one or more cows (or goats), including the milk fat, milk solids–not–fat, and fluid carrier portions, each to be computed and accounted for separately.
3. beverage which is approximately 87% water and 13% solids.

- **milk alkali syndrome**

a condition caused by excessive oral intake of calcium containing foods and antacids.

- **milk collection**

collection of milk from more than one farmer to a collection point or centre.

- **milk composition**

average composition of dairy cow milk includes the following constituents: 87% water, 3.25% protein, 2.5% casein, 5% sugar (lactose), 3.7% fat and 0.8% minerals and vitamins.

- **milk cooling**

cooling milk to increase keeping time.

- **milk equivalent**

the quantity of milk, as produced, required to furnish the milk solids in manufactured dairy products. For example, approximately 10 and 20 pounds of milk are required to manufacture 1 pound of Cheddar cheese and butter, respectively.

- **milk equivalent, milk fat basis**

pounds of milk containing a specified percentage of milk fat (usually 3.67 percent) needed to provide the milk fat contained in a given amount of a dairy product.

It is the traditional and most commonly used m.e. measure.

- **milk equivalent, total solids basis (m.e., t.s.)**
  like the milk equivalent milkfat basis except a formula is used to determine the total solids (milkfat and nonfat solids) contained in a given dairy product and the amount of standard test milk necessary to provide that amount of total solids.

- **milk fat**
  a complex mixture of triglycerides containing numerous fatty acids.

- **milk fever**
  see **parturient paresis**.

- **milk handlers**
  processors or dealers of milk who commonly purchase raw milk and sell pasteurised milk and milk products.

- **milk house**
  the area near a milking parlour where the bulk milk tank, cleaning units and equipment are located.

- **milk house waste**
  water having been used in cleaning the milking equipment and washing the parlour.

- **milk hygiene**
  making sure a milk product is clean and safe for consumption.

- **milk lipids - chemical properties**
  the fat content of milk is of economic importance because milk is sold on the basis of fat. Milk fatty acids originate either from microbial activity in the rumen, and transported to the secretory cells via the blood and lymph, or from synthesis in the secretory cells. The main milk lipids are a class called triglycerides which are comprised of a glycerol backbone binding up to three different fatty acids. The fatty acids are composed of a hydrocarbon chain and a carboxyl group. The major fatty acids found in milk are:
  Long chain
  C14 - myristic 11%
  C16 - palmitic 26%
  C18 - stearic 10%
  C18:1 - oleic 20%
  Short chain (11%)
  C4 - butyric*

C6 - caproic
C8 - caprylic
C10 - capric
* butyric fatty acid is specific for milk fat of ruminant animals and is resposible for the rancid flavour when it is cleaved from glycerol by lipase action.

Saturated fatty acids (no double bonds), such as myristic, palmitic, and stearic make up two thirds of milk fatty acids. Oleic acid is the most abundant unsaturated fatty acid in milk with one double bond. While the cis form of geometric isomer is the most common found in nature, approximately 5% of all unsaturated bonds are in the trans position as a result of rumen hydrogenation.

Lipid structures: 17 KB

Triglycerides account for 98.3% of milk fat. The distribution of fatty acids on the triglyceride chain, while there are hundreds of different combinations, is not random. The fatty acid pattern is important when determining the physical properties of the lipids. In general, the SN1 position binds mostly longer carbon length fatty acids, and the SN3 position binds mostly shorter carbon length and unsaturated fatty acids. For example:

C4 - 97% in SN3
C6 - 84% in SN3
C18 - 58% in SN1

The small amounts of mono-, diglycerides and free fatty acids in fresh milk may be a product of early lipolysis or simply incomplete synthesis. Other classes of lipids include phospholipids (0.8%), which are mainly associated with the fat globule membrane, and cholesterol (0.3%), which is mostly located in the fat globule core.

- **milk lipids - functional properties**

like all fats, milk fat provides lubrication. They impart a creamy mouth feel as opposed to a dry texture. Butter flavour is unique and is derived from low levels of short chain fatty acids. If too many short chain fatty acids are hydrolysed (separated) from the triglycerides, however, the product will taste rancid. Butter fat also acts as a reservoir for other flavours, especially in aged cheese. Fat globules produce a 'shortening' effect in cheese by keeping the protein matrix extended to give a soft texture. Fat substitutes are designed to mimic the globular property of milk fat. The spreadable range of butter fat is 16-24° C. Unfortunately butter is not spreadable at refrigeration temperatures. Milk fat provides energy (1g = 9 cal.), and nutrients (essential fatty acids, fat soluble vitamins).

- **milk lipids - physical properties**
  the physical properties of milk fat can be summarised as follows –
  density at 20° C is 915 kg m(-3)*
  refractive index (589 nm) is 1.462 which decreases with increasing temperature
  solubility of water in fat is 0.14% (w/w) at 20° C and increases with increasing temperature
  thermal conductivity is about 0.17 J m(-1) s(-1) K(-1) at 20° C
  specific heat at 40° C is about 2.1kJ kg(-1) K(-1)
  electrical conductivity is <10(-12) ohm(-1) cm(-1)
  dielectric constant is about 3.1
  *the brackets around numbers denote superscript.
  At room temperature, the lipids are solid, therefore, are correctly referred to as 'fat' as opposed to 'oil' which is liquid at room temperature. The melting points of individual triglycerides ranges from -75° C for tributyric glycerol to 72° C for tristearin. However, the final melting point of milk fat is at 37°C because higher melting triglycerides dissolve in the liquid fat. This temperature is significant because 37°C is the body temperature of the cow and the milk would need to be liquid at this temperature.
  The melting curves of milk fat are complicated by the diverse lipid composition:
  trans unstauration increases melting points
  odd-numbered and branched chains decrease melting points
  Crystallisation of milk fat largely determines the physical stability of the fat globule and the consistency of high-fat dairy products.

but crystal behaviour is also complicated by the wide range of different triglycerides. There are four forms that milk fat crystals can occur in; alpha, ß, ß ' 1, and ß ' 2, however, the alpha form is the least stable and is rarely observed in slowly cooled fat.

- **milk metre**
a device between the cluster or milking unit and the milk line for measuring a cow's milk yield in either mass or volume.

- **milk preservation**
increasing keeping time of milk. See **bronopol**.

- **milk processing**
processing of raw milk into milk products.

- **milk production map**
map of all milk production related items for participatory analysis.

- **milk protein fractionation**
the nitrogen content of milk is distributed among caseins (76%), whey proteins (18%), and non-protein nitrogen (NPN) (6%). This does not include the minor proteins that are associated with the FGM.

- **milk proteins**
1. casein and whey proteins.
2. the primary structure of proteins consists of a polypeptide chain of amino acids residues joined together by peptide linkages, which may also be cross-linked by disulphide bridges. Amino acids contain both a weakly basic amino group, and a weakly acid carboxyl group both connected to a hydrocarbon chain, which is unique to different amino acids. The three dimensional organisation of proteins, or conformation, also involves secondary, tertiary, and quaternary structures. The secondary structure refers to the spatial arrangement of amino acid residues that are near one another in the linear sequence. The alpha helix and ß -pleated sheat are examples of secondary structures arising from regular and periodic steric relationships. The

tertiary structure refers to the spatial arrangement of amino acid residues that are far apart in the linear sequence, giving rise to further coiling and folding. If the protein is tightly coiled and folded into a somewhat spherical shape, it is called a globular protein. If the protein consists of long polypeptide chains, which are intermolecularly linked, they are called fibrous proteins. Quaternary structure occurs when proteins with two or more polypeptide chain subunits are associated.

- **milk sampling**
  taking small amounts of milk for analysis later.
- **milk serum**
  the nonfat components of milk.
- **milk shed**
  a geographical area in which milk is produced for a given milk-marketing area.
- **milk soil**
  residue on equipment, utensils or containers after - emptying.
- **milk solids**
  1. the solids portion dissolved in milk made up of 4% fat and 9% solids non-fat - including milk proteins and lactose.
  2. combined fat and protein yield of milk.
- **milk solids-not-fat (SNF)**
  the solids portion of the milk minus the fat component represents about 8.5 to 9.2 percent of the total milk solids.
- **milk testing**
  checking the hygiene and composition of milk.
- **milk volume**
  the amount of milk produced.
- **milk well**
  the opening in the abdominal wall through which milk veins (subcutaneous abdominal veins) enter to join the vena cava and return blood from udder to heart.
- **milkfat structure - fat globules**
  more than 95% of the total milk lipid is in the form of a globule ranging in size from 0.1 to 15 um in diameter. A thin membrane, 8 to 10 nm

in thickness, whose properties are completely different from both milk fat and plasma, covers these liquid fat droplets. The native fat globule membrane (FGM) is comprised of apical plasma membrane of the secretory cell, which continually envelops the lipid droplets as they pass into the lumen. The major components of the native FGM, therefore, are protein and phospholipids. The phospholipids are involved in the oxidation of milk. There may be some rearrangement of the membrane after release into the lumen as amphiphilic substances from the plasma adsorb onto the fat globule and parts of the membrane dissolve into either the globule core or the serum. The FGM decreases the lipid-serum interface to very low values, 1 to 2.5 mN/m, preventing the globules from immediate flocculation and coalescence, as well as protecting them from enzymatic action.

It is well known that if raw milk or cream is left to stand, it will separate. **Stokes' Law** predicts that fat globules will cream due to the differences in densities between the fat and plasma phases of milk. However, in cold raw milk, creaming takes place faster than is predicted from this fact alone. IgM, an immunoglobulin in milk, forms a complex with lipoproteins. This complex, known as cryoglobulin precipitates onto the fat globules and causes flocculation. This is known as cold agglutination. As fat globules cluster, the speed of rising increases and sweeps up the smaller globules with them. The cream layer forms very rapidly, within 20 to 30 min., in cold milk.

Homogenisation of milk prevents this creaming by decreasing the diameter and size distribution of the fat globules, causing the speed of rise to be similar for the majority of globules. As well, homogenisation causes the formation of a recombined membrane, which is much similar in density to the continuous phase.

Recombined membranes are very different than native FGM. Processing steps such as homogenisation decreases the average diameter of fat globule and significantly increases the surface area. Some of the native FGM will remain adsorbed but there is no longer enough of it to cover all of the newly created surface area. Immediately after disruption of the fat globule, the surface tension raises to a high level of 15 mN/m and amphiphilic molecules in the plasma quickly adsorb to the lipid droplet to lower this value. The adsorbed layers consist mainly of serum proteins and casein micelles.

Electron Micrograph of Homogenised Milk fat Globules 77 KB

Fat Destabilisation

While homogenisation is the principal method for achieving stabilisation of the fat emulsion in milk, fat destabilisation is necessary for structure formation in butter, whipping cream and ice cream. Fat destabilisation refers to the process of clustering and clumping (partial coalescence) of the fat globules, which leads to the development of a continuous internal fat network or matrix structure in the product. Fat destabilisation (sometimes 'fat agglomeration') is a general term that describes the summation of several different phenomena. These include:

Coalescence:

an irreversible increase in the size of fat globules and a loss of identity of the coalescing globules,

Flocculation:

a reversible (with minor energy input) agglomeration/clustering of fat globules with no loss of identity of the globules in the floc, the fat globules that flocculate. They can be easily redis-persed if they are held together by weak forces, or they might be harder to redisperse to they share part of their interfacial layers.

Partial coalescence:

an irreversible agglomeration/clustering of fat globules, held together by a combination of fat crystals and liquid fat, and a retention of identity of individual globules as long as the crystal structure is maintained (i.e., temperature dependent, once the crystals melt, the cluster coalesces). They usually come together in a shear field, as in whipping, and it is envisioned that the crystals at the surface of the droplets are responsible for causing colliding globules to stick together, while the liquid fat partially flows between they and acts as the 'cement'. Partial coalescence dominates structure formation in whipped, aerated dairy emulsions and it should be emphasised that crystals within the emulsion droplets are responsible for its occurrence.

- **milking pit**
  a sunken area that houses both the milker and some milking equipment during milking. This places the milker at shoulder level with udders and reduces physical demands.
- **milkshed**
  area wherein the producers are located who supply the plants regulated under a given marketing order. May also refer to the area in which competing plants draw their milk supply. The term can also apply to the milk supply of a single plant.
- **milkstone**
  residue on equipment, utensils, or containers used for handling milk. It is characterised by a high mineral content, especially calcium. It is normally soluble in acid.
- **mill lodge**
  an articificial pond created in order to provide water for an industrial process. In Lancashire, many examples are associated with textile mills and some used to provide water to drive water wheels.
- **mill pond**
  see **mill lodge**.
- **millilitre**
  one thousandth of a litre.
- **millstone grit**
  coarse-grained sandstone that divides carboniferous limestones from coal-bearing rock strata.
- **mineral**
  a nutrient required to maintain health.

- **minerals**

nutrients required by humans in small amounts for body processes such as maintenance, structure and regulation. Can't be made in the body and must be obtained from foods.

All 22 minerals considered to be essential to the human diet are present in milk. These include three families of salts Sodium (Na), Potassium (K) and Chloride (Cl)

These free ions are negatively correlated to lactose to maintain osmotic equilibrium of milk with blood.

Calcium (Ca), Magnesium (Mg), Inorganic Phosphorous (P(i)), and Citrate. This group consists of 2/3 of the Ca, 1/3 of the Mg, 1/2 of the P(i), and less than 1/10 of the citrate in colloidal (no diffusible) form and present in the casein micelle.

Diffusible salts of Ca, Mg, citrate, and phosphate.

These salts are very pH dependent and contribute to the overall acid-base equilibrium of milk.

- **mire**

plant communities that develop on waterlogged land.

- **misbranded**

this term applies to any carcass, part thereof, meat or meat food product under one or more of the following circumstances: 1) If its labelling is false or misleading in any particular 2) If it is offered for sale under the name of another food 3) If it is an imitation of another food, unless its label bears, in type of uniform size and prominence, the word 'imitation' and immediately thereafter, the name of the food imitated 4) If its container is so made, formed, or filled as to be misleading 5) If in a package or other container unless it bears a label showing: a.) the name and place of business of the manufacturer, packer, or distributor and b.) an accurate statement of the quantity of the contents in terms of weight, measure, or numerical count, except as otherwise provided.

- **mitolactol**

an anticancer drug that makes tumour cells more sensitive to radiation therapy.

- **mitosis**
process of cell reproduction whereby the daughter cells are identical in chromosome number to the parent cells. Compare Meiosis.

- **mitotane**
a drug that increases the amount of chemotherapy drugs that can accumulate in tumour cells.

- **mixed farming**
agriculture where arable and livestock farming are practiced together on the same farm holding.

- **mobile dune**
a coastal sand dune that is becoming stabilised by vegetation but which is still prone to significant erosion by wind erosion. Compare with fixed dune.

- **modified**
changed. When applied to 'BOGS', it means that the natural hydrology of the bog has been altered artificially. For example, PEAT cutting or drainage of peatlands will modify lowland raised bogs.

- **modified atmosphere packaging (MAP)**
packaging method in which a combination of gases such as oxygen, carbon dioxide and nitrogen is introduced into the package at the time of closure. The purpose is to extend shelf life of the product packaged.

- **modified food starch**
starch that has been chemically altered to improve its thickening properties. Before the starch is modified, it is separated from the protein through isolation techniques, therefore, the source of the starch used is not required on the label.

- **mohair**
the long, lustrous fleece covering Angora goats, used for special kinds of cloth.

- **molasses**
means a thick syrup which is a byproduct of processing sugar beets or sugarcane or of refining raw cane sugar.

- **molecular biology**
general term referring to study of the structure and function of proteins and nucleic acids, may be used as a synonym for genetic

engineering or recombinant DNA techniques.

- **molecular genetics**
  study of how genes function to control cellular activities.

- **mollusc**
  a member of a varied group of invertebrates that secrete calcium carbonate to form shells. The shells may be external and obvious (as in snails and mussels) or internal and small (as in slugs).

- **molluscicide**
  a chemical that kills molluscs. Usually applied to chemicals (e.g. slug pellets) intended to reduce populations of slugs and snails.

- **mollusk**
  invertebrate animals with soft body coverings and shells of one to eight parts or sections. This group includes clams, mussels, oysters, abalone and snails.

- **monoclonal antibodies**
  antibodies made from single cells that recognise one chemical component on a foreign target molecule.

- **monoclonal antibody**
  laboratory produced substances that are directed against cancer cells and can be used to deliver drugs, toxins, or radioactive material directly to the cancer cells. There are many monoclonal antibodies used in cancer therapy and each one recognises a different protein on a variety of cancer cells. Monoclonal antibodies include A33,17-la, B3, BR55-2, BR96, C225, CC-49, Di-dgA-RFB4, Mik-beta-1, NR-LU-10, PM-81, Prost 30, and rituximab.

- **monoclonal antibody (MAB)**
  highly specific, purified antibody that is derived from only one clone of cells and recognises only one antigen. See **Hybridoma.**

- **monocytes**
  one of three types of white blood cells. Monocytes are precursors to macrophages.

- **monogastric**
  having only one stomach or stomach compartment, as does man, dogs and swine.

- **monosodium glutamate (MSG)**
  1. msg is a flavour enhancer. It comes from a common amino acid, glutamic acid, and must

be declared as monosodium glutamate on meat and poultry labels.

2. a salt of glutamic acid used for the treatment of hepatic coma and the enhancement of the flavour of foods.

- **montane**
the zone above the height at which trees naturally grow.

- **montanide isa-51**
a drug used in vaccine therapy to stimulates the immune system.

- **moorland**
unenclosed land in the uplands that supports upland heath, blanket bog and upland grassland.

- **mosaic**
an interlinking mix of habitats covering a particular area. For example, a mosaic of scrub, grassland and woodland.

- **moss**
a group of bryophytes, the members of which are differentiated into stem and leaves (Compare with liverworts). An abbreviated form of mossland.

- **mounting**
the act of an animal raising its front legs onto the back of another animal in preparation for mating or before ejaculation.

- **mucositis**
inflammation of the lining of the digestive system, often seen as sores in the mouth, that is a complication of some cancer therapies.

- **muley**
a polled cow.

- **multidrug resistance**
adaptation of tumour cells to chemotherapy by eliminating the drugs faster or breaking them down so that they are no longer effective.

- **multidrug resistance inhibition**
treatment used to suppress resistance of cells to chemotherapy.

- **multigenic**
of hereditary characteristics, one that is specified by several genes.

- **multimodality treatment**
therapy that combines more than one method of treatment, such as chemotherapy and radiotherapy.

- **multiple myeloma**
  cancer of the bone marrow in multiple sites.
- **mummified foetus**
  a shriveled foetus that has remained in the uterus instead of being aborted or expelled. Fluids from the fetus have been partially reabsorbed by the mother.
- **mutagen**
  a substance that induces mutations.
- **mutant**
  a cell that manifests new characteristics due to a change in its DNA.
- **mutation**
  1. a change in the genetic material of a cell.
  2. a change in a gene that causes it to make a different product.
- **mutton**
  meat from older sheep that were once part of a breeding herd.
- **mycophenolate mofetil**
  an investigational drug that is being studied for its effectiveness in preventing graft-versus-host disease and autoimmune disorders.
- **mycoplasma**
  an organism capable of causing mastitis.
- **mycosis fungoides**
  a long-term, rapidly developing form of cutaneous T-cell lymphoma, a cancerous disease that affects the skin. It commonly results in infections or tumours of the skin that can spread to the lymph nodes or other organs such as the spleen, liver or lungs
- **myelodysplastic syndrome**
  disease in which the bone marrow does not function normally and not enough blood cells are made. Also called pre-leukemia or 'smoldering leukemia'.
- **myeloma**
  1. a type of tumour cell that is used in monoclonal antibody technology to form hybridomas.
  2. cancer found in bone marrow cells.
- **myeloproliferative disorders**
  diseases in which too many blood cells are made in the bone marrow. Myeloproliferative disorders include myelofibrosis.

- **myelosuppressive therapy**
  inhibition of blood cell production.
- **n-acetyl cysteine**
  an antioxidant drug that may keep cancer cells from developing or prevent the growth of existing cancer.
- **native species**
  species that occur naturally in an area and therefore one that has not been introduced by humans either accidentally or intentionally.
- **natural**
  a product containing no artificial ingredient or added colour that is only minimally processed (a process which does not fundamentally alter the raw product) may be labelled natural. The label must explain the use of the term natural (such as no added colourings or artificial ingredients, minimally processed.)
- **natural active immunity**
  immunity that is established after the occurrence of a disease.
- **natural cheese**
  product that is a single variety of cheese made from milk. The length of time natural cheese is allowed to ripen determines its flavor (mild, medium, or sharp). Brie, Mozzarella, Cheddar and Parmesan are all natural cheeses.

- **natural killer cell (NK)**
  a type of leukocyte that attacks cancerous or virus-infected cells without previous exposure to the antigen. NK cell activity is stimulated by interferon.
- **natural passive immunity**
  immunity conferred by the mother on the foetus or newborn.
- **natural regeneration**
  natural regrowth (e.g. new branches growing from the base of a coppiced tree.)
- **natural service in farm animals,**
  it means to allow natural mating, as opposed to artificial insemination.

▪ **nectar**
the sugary fluid secreted by the nectaries of flowers to attract the animals that cause pollination.

▪ **nectary**
a glandular structure in a flower that secretes nectar.

▪ **needle biopsy**
the removal of tumour tissue or fluid from a tumour for microscopic examination. This procedure is sometimes called fine needle aspiration.

▪ **negative axillary lymph nodes**
nodes in the armpit that have been examined and found to be free of cancer.

▪ **nemertean**
member of a group of animals known as 'proboscis' or 'ribbon' worms. These worm-like organisms possess a thin 'proboscis' organ that they can extrude to capture food or for locomotion.

▪ **neoadjuvant therapy**
therapy given before the primary treatment to improve the effectiveness of the treatment. Neoadjuvant therapy can be chemotherapy, radiation therapy or hormone therapy.

▪ **neonatal**
pertaining to a newborn animal.

▪ **neoplasia**
abnormal and uncontrolled cell growth.

▪ **neoplastic meningitis**
cancer cells that have spread to the tissue that covers the brain and spinal cord.

▪ **nephrectomy**
surgical removal of a kidney.

▪ **nephrocalcinosis**
a condition where one or both kidneys is full of stones located within the renal tissue. Usually associated with an underlying medical problem like Renal Tubular Acidosis.

▪ **nephrolithiasis**
condition in which one or more stones is located in a kidney.

▪ **nephrolithotomy**
standard surgical procedure in which the kidney is opened and a stone is removed.

▪ **nephrologist**
a medical specialist primarily in kidney problems. Nephrologists are usually

involved in cases of renal failure and dialysis, but many are very interested and knowledgeable about kidney stone disease and prevention. They are not surgeons.

- **nephrology**
  the medical specialty that deals with kidney diseases.
- **nephron**
  the microscopic functional unit of the kidney. Each kidney is made up of thousands of nephrons.

- **neutral detergent fibre (NDF)**
  a measurement of fiber after digesting in a non-acidic, non-alkaline detergent as an aid in determining quality of forages. Contains the fibers in ADF, plus hemicellulose.

- **neutropenia**
  an abnormal decrease in the number of certain white blood cells in the blood.

- **nicotinamide**
  a form of vitamin B3. It acts to increase the effect of radiation therapy on tumour cells.

- **nimodipine**
  belongs to a group of drugs known as calcium channel blockers. It is being investigated for use with chemotherapy to prevent or overcome drug resistance and improve response to chemotherapy.

- **nitrification**
  the biochemical oxidation of ammonium to nitrate, predominantly by autotrophic bacteria.

- **nitrification**
  the biochemical transformation by oxidation of ammonium (NH4) to nitrite (NO2) or to nitrate (NO3).

- **nitrogen balance**
  nitrogen in the food consumed minus nitrogen in feces and nitrogen in urine (nitrogen retention).

- **nitrogen fixation**
  a biological process (usually associated with plants) whereby certain bacteria convert nitrogen in the air to ammonia, thus forming a nutrient essential for growth.

- **nitrogen fixing**
  conversion of atmospheric nitrogen gas to a chemically combined form, ammonia, which is essential to plant growth. Some microorganisms can 'fix' nitrogen.

- **nitrogen-free extract (NFE)**
  consisting of carbohydrates, sugars, starches, and a major portion of materials classed as hemicellulose in feeds. When crude protein, fat, water, ash, and fiber are added and the sum is subtracted from 100, the difference is NFE.

- **no roll**
  the term 'No Roll' is permitted on marking devices and

labels for single ingredient red meats (carcasses, primal and retail cuts) provided the term is not accompanied with an official grade name (e.g., 'No Roll Choice').

- **nocturnal**
  active at night.
- **node-negative**
  cancer has not spread to the lymph nodes.
- **nominated semen**
  livestock Improvement's artificial breeding option, Cheqmate and Globull, which allows farmers to choose which specific sires they want to use in their herd and on specific cows.
- **noncoding DNA**
  DNA that does not encode any product (RNA or protein). The majority of the DNA in plants and animals is non-coding.
- **nonfat dry milk**
  product obtained by removing water from pasteurised skim milk. It has all the protein, minerals and vitamins of nonfat milk and is fortified with vitamins A and D- It takes 100 pounds of skim milk to make 8 pounds of nonfat dry milk.
- **non-hodgkin's lymphoma**
  a group of malignant cancers of the immune system, including acute lymphoblastic leukemia, B cell lymphoma, Burkitt's lymphoma, diffuse cell lymphoma, follicularlymphoma, immunoblastic large cell lymphoma, lymphoblastic lymphoma, mantle cell lymphoma, mycosis fungoides, post-transplantation lympho-proliferative disorder, small noncleaved cell lymphoma and T-cell lymphoma.
- **nonmetastatic**
  cancer that has not spread from the primary site of origin to other sites in the body.
- **non-point source**
  entry of effluent into a water body in a diffuse manner so there is no definite point of entry.
- **non–pool milk**
  milk received at non-pool plants.
- **non–pool plants**
  plants receiving, processing or manufacturing dairy products but which do not

participate in the milk pooling program.

- **non-renewable**
a resource that is in effect finite because it is not replaced within normal human timescales. For example, oil is non- renewable because it takes millions of years to form and accumulate.

- **non-return**
the breeding efficiency of bulls expressed as the percentage of cows that conceive on first service. Also called non-return rate.

- **non-return dip cup**
a dip cup that does not allow the liquid to re-enter and potentially contaminate the storage container.

- **NPK**
nitrogen, phosphorous and potassium used in describing the composition of a fertiliser.

- **NPN nonprotein nitrogen (e.g., urea).**
used by rumen microorganisms to synthesise protein.

- **nr-lu-10 antigen**
a protein found on the surface of some cancers.

- **NSAIDS**
nonsteroidal anti-inflammatory drugs (NSAIDS) decrease the symptoms of swelling, pain and redness.

- **nuclease**
an enzyme that, by cleaving chemical bonds, breaks down nucleic acids into their constituent nucleotides.

- **nucleic acids**
large molecules, generally found in the cell's nucleus and/or cytoplasm, that are made up of nucleotides. The two most common nucleic acids are DNA and RNA.

- **nucleotides**
the building blocks of nucleic acids. Each nucleotide is composed of sugar, phosphate and one of four nitrogen bases. The sugar in DNA is deoxyribose and RNA's sugar is ribose. The sequence of the bases within the nucleic acid determines the sequence of amino acids in a protein.

- **nucleus**
the structure within eukaryotic cells that contains chromosomal DNA.

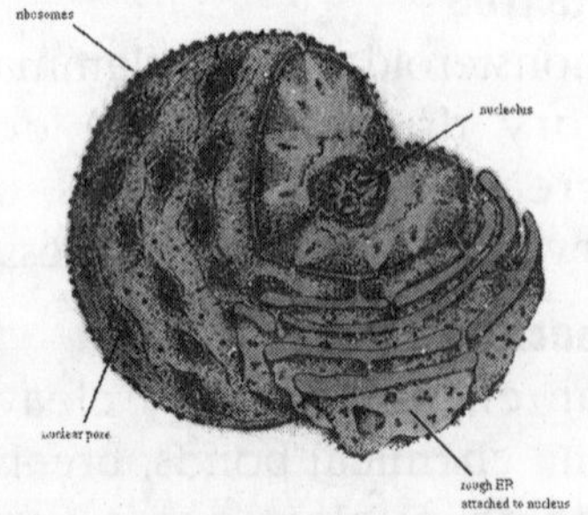

■ **nurse cow**
a dairy cow used to supply milk for nursing calves, commonly other than her own.

■ **nutraceutical**
food having pharmaceutical benefits.

■ **nutrient**
1. a substance that provides nourishment.
2. chemicals in foods that are required for metabolism, growth and repair or provide energy. Protein, carbohydrate, fat, vitamins, minerals and water are all nutrients.

■ **observation**
the patient's condition is closely monitored, but treatment is withheld until symptoms appear or change.

■ **octreotide**
a hormone-like drug. Also known as Sandostatin.

■ **oestrus**
see **estrus.**

■ **off feed**
having ceased eating; without a healthy and normal appetite.

■ **offspring**
the young of an animal.

■ **oligochaete**
the group of animals that includes terrestrial earthworms and freshwater 'blood worms' (also called 'Tubifex' worms).

■ **oligonucleotide**
a polymer consisting of a small number (about two to 10) of nucleotides.

■ **oligotrophic**
nutrient poor (not necessarily base poor).

■ **oltipraz**
an investigational cancer prevention drug.

■ **omasum**
the third division of the stomach of ruminants. Also called manifold, manyplies and psalterium.

- **ombrotrophic**
  supplied solely by water derived from the atmosphere (rain, snow, fog etc.).

- **omega-3 fatty acid**
  a type of fat obtained in the diet that is involved in immunity.

- **omeprazole**
  a drug that inhibits gastric acid secretion.

- **on the hoof**
  designating a living meat animal.

- **oncogene**
  gene thought to be capable of producing cancer.

- **oncogenic**
  cancer causing.

- **oncology**
  study of tumours.

- **ongoing**
  in progress, developing, continuing. In terms of the Action Tables in this Plan it means an action that, by its nature, needs to kept under constant periodic review.

- **open**
  a term commonly used for non-pregnant farm mammals.

- **open barns/tromp sheds/loose housing**
  open spaced shelter in which cattle are free to move about or rest wherever they might prefer, usually on a pack of bedding and manure.

- **openness**
  a quality in dairy cattle associated with length of body and width and length of rib.

- **operator gene**
  a region of the chromosome, adjacent to the operon, where a repressor protein binds to prevent transcription of the operon.

- **operon**
  sequence of genes responsible for synthesising the enzymes needed for biosynthesis of a molecule. An operon is controlled by an operator gene and a repressor gene.

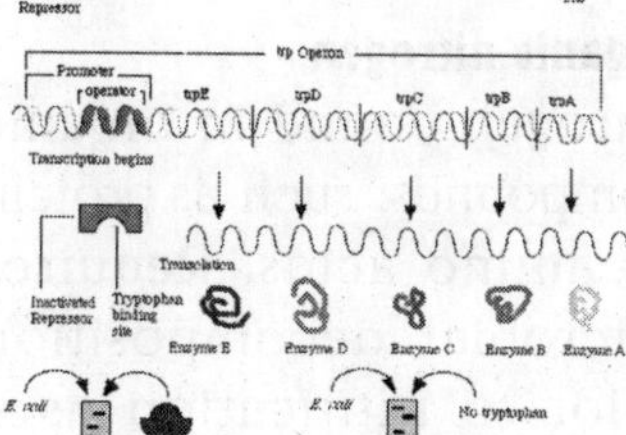

- **opioid**
  a class of strong pain medicines derived from Opium. Morphine and Codeine would be examples.
- **order**
  a unit in the classification of plants and animals.
- **organic**
  Of, or produced by animals and plants. When applied to chemical substances means compounds containing carbon combined with hydrogen (and often with nitrogen and/or oxygen).
- **organic compound**
  a compound containing carbon.
- **organic material**
  substances containing plant or animal substance. In the context of milking equipment this usually refers to manure.
- **organic nitrogen**
  nitrogen bound in organic compounds, such as protein or amino acids. Requires microbial decomposition prior to nitification as a plant nutrient.
- **organic pollution/enrichment**
  artificial elevation of the levels of organic chemicals in soil or water.
- **organism**
  a living animal or plant.

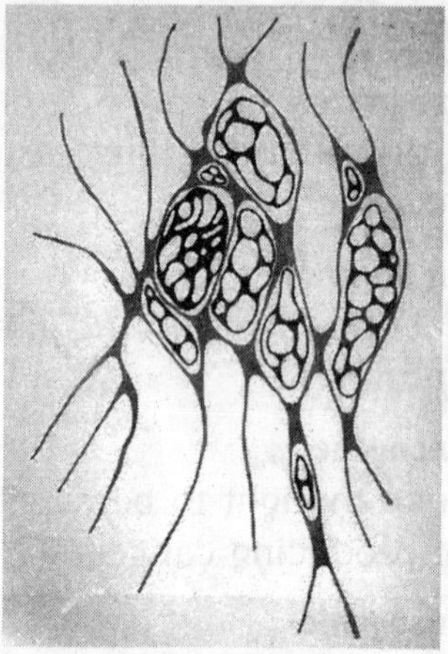

- **organochlorine**
  an organic compound that contains chlorine. These compounds were artificially produced for use as insecticide. Once widely used in agriculture there are now restrictions on their application since it was found that they were accumulating in vertebrates and poisoning them.
- **organoleptic tests**
  tests based on taste, smell or visual observations.

- **osmotic pressure**
  a chemical force caused by a concentration gradient. It is a colligative property and the principle behind membrane processing.

- **osteolytic**
  causing the breakdown of bone.

- **osteoporosis**
  condition in which the bones are weak because of loss of calcium. This often occurs in people with forced immobilisation, astronauts, women after menopause and in patients with hyperparathyroidism. Usually treated with supplemental calcium and either estrogen, Fosamax, calcitonin, exercise or some combination of the above.

- **out of**
  terminology for mothered by in animal breeding.

- **outcross**
  mating an individual to another in the same breed who is not closely related to it.

- **outlier**
  a population that occurs outside the main range of the species.

- **ovariectomy**
  the surgical removal of an ovary.

- **ovaries**
  female reproductive organs that produce eggs.

- **over the counter drugs**
  medications available without prescription.

- **overconditioned**
  an animal with excess flesh and patchy fat deposits, especially over the shoulders, back, hips and thighs.

- **overexpress**
  when a cell has an excess of a particular protein on its surface.

- **overgrazing**
  grazing that, from the conservation point of view, is so intense that it is damaging or degrading the habitat.

- **overhead wires**
  wires that run above a paddock onto which a bull is attached by a chain for free grazing.

- **overwinter**
  survive over the winter period.

■ **oxalate**
a chemical compound that is found in most calcium kidney stones. It binds strongly with calcium. Dietary sources of oxalate include tea, chocolate, green leafy vegetables, nuts, tomatoes and some cola soft drinks.

■ **oxaliplatin**
an investigational platinum analogue anticancer drug.

■ **oxalobacter**
the name of a normal bacteria of the intestinal tract that has the unique ability to digest oxalate. Currently used only experimentally, but has great potential as a future therapy for oxalate problems.

■ **oxygen**
the reactive gas that makes up 20% of the air we breathe.

■ **oxytocin**
1. a naturally secreted hormone that is important in milk letdown and the contraction of the smooth uterine muscles during the birthing process.
2. a naturally secreted hormone that is important in milk letdown and the contraction of smooth uterine muscles during parturition.

■ **p-30 protein**
a chemotherapy agent currently being tested to determine if its use with tamoxifen will be effective treatment for tumours that are resistant to chemotherapy. Also known as Onconase.

■ **p53 gene**
a gene that has been found to be mutated in many types of cancer. Also called a tumour-suppressor protein.

■ **packaged milk**
class 1 or beverage fluid milk products as packaged in final form and distinct from bulk milk.

■ **packer**
a firm that slaughters or slaughters and processes livestock or poultry.

■ **paclitaxel**
one of the chemotherapy agents called taxanes that block cell division. Also called Taxol.

■ **paddock**
an area of grass surrounded by fences, may also be called a field.

■ **paddocks**
subdivision of a pasture designed to provide short-duration grazing followed by an appropriate (related to species, soil type and weather conditions) rest period for regrowth and stand maintenance.

■ **paint sticks**
contain liquid or chalky paint used for marking treated cows.

■ **pala**
an anticancer drug than enhances the effectiveness of the chemotherapy drug fluorouracil.

■ **palaearctic**
a region of the earth defined in terms of its plants and animals that covers Europe, Asia and North Africa.

■ **palliative therapy**
treatment given to relieve discomfort or pain caused by advanced cancer.

■ **palpation**
examination by touch.

■ **pamidronate**
belongs to the group of drugs known as bisphosphonates that are used as treatment for hypercalcemia (abnormally high levels of calcium in the blood) and for cancer that has spread to the bone (bone metastases).

■ **parakeratosis**
any abnormality of the stratum corneum (horny layer of epidermis) of skin, especially a condition caused by edema between the cells, which prevents the formation of keratin.

■ **parallel milking parlour**
a raised milking area or platform where the cow stands perpendicular to the operator and milking units are attached between the rear legs. Also called side-by-side.

■ **parallel parlour**
a raised milking area or platform where the cow stands

perpendicular to the operator and milking units are attached between the rear legs. This may also be referred to as a side-by-side.

- **parallel**
  swing and rotary.

- **parapox**
  a type of virus.

- **parasites**
  organisms that derive nourishment and protection from other living organisms known as hosts. They may be transmitted from animals to humans, from humans to humans, or from humans to animals. Several parasites have emerged as significant causes of foodborne and waterborne disease. These organisms live and reproduce within the tissues and organs of infected human and animal hosts and are often excreted in faeces. Some common parasites are Giardia duodenalis, Cryptosporidium parvum, Cyclospora cayetanensis, Toxoplasma gondii, Trichinella spiralis, Taenia saginata (beef tapeworm), and Taenia solium (pork tapeworm).

- **parathyroid hormone**
  the name of the hormone that comes from the parathyroid glands. This hormone increases blood calcium. To do this, it will increase calcium absorption from the digestive tract and take calcium away from the bones and teeth. Abnormally high levels of this hormone produce a disease called Hyperparathyroidism.

- **parenteral**
  1. means 'Into the Body' Usually refers to any type of injectable medication.
  2. administration by injection, not through the digestive tract (e.g., subcutaneous, intramuscular, intrasternal, intravenous).

- **parity price**
  a level of farm prices intended to provide the same approximate purchasing power for farmers as enjoyed during a preceding base period.

- **parlour**
  the specialised area on the dairy farm where milking is

performed. Parlors come in many types: flat barn, walk-through, herringbone.

- **paroxetine hydrochloride**
  an antidepressant drug. Also called Paxil

- **parrot mouth**
  a malformed mouth of an animal in which the upper jaw abnormally protrudes beyond the lower.

- **partial migrant**
  a species in which some individuals migrate but not all. For example, some British song thrushes remain where they are in winter others move to the European mainland.

- **partial remission**
  the shrinking, but not complete disappearance, of a tumour in response to therapy. Also called partial response.

- **partial response**
  when tumours shrink as a result of therapy but do not completely disappear.

- **partially defatted (beef or pork) fatty tissue**
  these are by-products produced from fatty trimmings containing less than 12% lean meat. These ingredients may be used in meat products in which by-products are acceptable.

- **participation**
  acknowledging ideas of all group members.

- **parturient paresis**
  a condition (observed especially in high producing dairy cows) characterised by a low blood calcium concentration which results in partial to complete paralysis soon after parturition called milk fever.

- **parturition**
  the act or process of giving birth to young.

- **passive immunity**
  immunity acquired from receiving preformed antibodies.

- **pasteurisation**
  1. destroying any potential pathogenic germs by heating milk at a minimum of 63°C for 30 min.
  2. under HTST (high temperature, short time) pasteurisation, milk is heated to 160° – 175° F for 15 to 25 seconds to destroy pathogenic and other undesirable micro-organisms

that may be found in milk. Ultra high temperature (UHT) pasteurisation occurs at 275° to 280° F for about 2 seconds.

3. process of heating a liquid to a high temperature and chilling it quickly to destroy harmful bacteria without causing a major chemical change to the liquid itself.
4. the process of destroying microorganisms that could disease. This is usually done by applying heat to food. Three processes used to pasteurise foods are flash pasteurisation, steam pasteurisation, and irradiation pasteurisation.
5. the process of heating milk to at least 62.8°C. (145°F.) and holding it at that temperature for not less than 30 minutes (holding method); to 71.7°C. (161°F.) for 15 seconds (HTST); to 88.3°C. (191°F.) for 1 second, to 95.5°C. (203.9°F.) for 0.05 second, or to 100°C. (212°F.) for 0.01 second (ultra pasteurisation).

■ **pastoral**
relating to pasture.

■ **pastoral farming**
grass based.

■ **pasture**
1. grassland used for grazing livestock.
2. plants, such as grass, grown for feeding or grazing animals. Also serves as a place to feed cattle and other livestock.
3. plants, as grass, grown for feeding or grazing animals. Also to feed cattle and other livestock on pasture.

■ **pathogen**
1. disease-causing agent.
2. a microorganism (bacteria, parasites, viruses, or fungi) that is infectious and causes disease.
3. a specific causative agent of disease.
4. any microorganism that produces disease (bacteria, viruses, yeasts, molds and parasites).

■ **pathologic fracture**
a broken bone caused by disease, often by metastases to the bone.

- **paunch (rumen)**
  the large, first stomach compartment of a ruminant.

- **PCA**
  literally 'Patient Controlled Analgesia'. This is a program where a machine dispenses a preset amount of pain medicine directly into a vein when the patient pushes a button.

- **pediculicide**
  an agent that destroys lice.

- **pedigree**
  a record of an animal's ancestors, usually only those of the five closest generations. A complete pedigree may include milk production, type classification and other information pertaining to dairy cattle.

- **peg-mgdf**
  a synthetic form of a protein that is normally made in the body to regulate the production of platelets.

- **peldesine**
  a chemotherapy drug used to treat T-cell lymphomas.

- **penclomedine**
  a chemotherapy drug that slows the growth of some cancers.

- **pendulous udder**
  1. a loosely attached udder.
  2. loosely attached mammary glands are collectively referred to as pendulous udders.

- **penis**
  male reproductive organ.

- **pentoxifylline**
  a drug used to prevent blood clotting and as an investigational treatment that may help reduce weight loss in cancer patients.

- **peptide**
  1. two or more amino acids joined by a linkage called a peptide bond.
  2. any compound consisting of two or more amino acids, the building blocks of proteins. Peptides are combined to make proteins.

- **peptide 946**
  a protein that causes white blood cells to recognise and destroy melanoma cells.

- **percentage**
  part of a whole amount.

- **percentile (%ile)**
  for bulls, percentile (%ILE) is a ranking based on the Active AI bull population, for

cows, %ILE is a ranking based on cows with recent lactations. With the Animal Model, PTA$ Protein is used to determine percentiles.

- **percutanecus**
literally means through the skin. Several different types of kidney stone procedures can be done by surgically creating a passage directly into the central, open part of the kidney and passing a variety of probes, graspers, stents, catheters and other devices through the passage and into the kidney. Very useful for large or complicated stones.

- **perennial**
a plant living for at least three seasons. Normally a perennial flowers in its second season and subsequently.

- **perfusion**
bathing an organ or tissue, or the blood vessels of an organ, with a fluid.

- **pericardial effusion**
presence of fluid inside the membrane that covers the heart.

- **perillyl alcohol**
an investigational drug that may inhibit the growth of some cancers.

- **peripheral blood**
circulating blood.

- **peripheral stem cell transplantation**
similar to bone marrow transplantation, peripheral stem cell transplantation is a method of replacing blood-forming cells destroyed by cancer treatment. Immature blood cells (stem cells) in the blood that are similar to those in the bone marrow are removed from the patient's blood before treatment. These cells are given back to the patient after treatment to help the bone marrow recover and continue producing healthy blood cells. Transplantation may be autologous (the patient's own blood cells saved earlier), allogeneic (blood cells from someone else) or syngeneic (blood cells from an identical twin).

- **peripheral stem cells**
immature cells from which blood cells that are found

circulating in the bloodstream develop.

- **perishable**
food that is subject to decay, spoilage, or bacteria unless it is properly refrigerated or frozen.

- **peritoneum**
the tissue that lines the abdominal wall and covers most of the organs in the abdomen. A peritoneal perfusion and an intraperitoneal infusion are methods of delivering fluids and drugs to tumours in the peritoneum.

- **permanent identification**
identification that stays with the animal for its lifetime and cannot be lost. Examples are tatoo, colour markings (sketch or photo) and hot or freeze brand.

- **permeable**
a body is said to be permeable to a substance if its allows the passage of the substance through itself.

- **persistency**
the quality of being persistent, as in the ability of lactating animals to maintain milk production over a period of time.

- **person**
means an individual, corporation, association, marketing or processing cooperative, joint stock company, estate or trust, or other legal entity.

- **pesticide**
chemical used to destroy pests.

- **pet**
pet (positron emission tomography) is an x-ray test to determine the activity and location of the tumour.

- **PFF**
protein fat free.

- **pH**
  1. a measure of acidity/alkalinity. A pH value of 1.0 is extremely acidic, 7.0 is 'neutral', while values above 7.0 are basic or alkaline.
  2. a measurement of the acid content of a liquid such as urine. A pH of 7 would be 'neutral' like water. Acids have a low pH while alkaline liquids and bases have high pH numbers. The usual range for living organisms is between 5.0 and 9.0.

3. lowering the pH leads to dissolution of calcium phosphate until, at the isoelectric point (pH 4.6), all phosphate is dissolved and the caseins precipitate.
4. pH is a measure of the activity of the hydronium ion (H_3O^+) which, according to the Debye-Huckel expression, is a function of the concentration of the hydronium ion [H_3O^+], the effective diameter of the hydrated ion and the ionic strength (μ m) of the solvent. For solutions of low ionic strength (μ m < 0.1) hydronium ion activity is nearly equivalent to [H_3O^+] which is normally abbreviated to [H^+]. Then, for a weak acid (HA) dissociating to H^+ and A^- with a dissociation constant, Ka and pKa equal to -log10 Ka, the most important relationships are defined the following two equations $Ka = [H^+] [A^-] / [HA]$

   $pH = \log 1 / [H^+] = pKa + \log [A^-] / [HA]$
5. a measure of the acidity or basicity of a solution.
6. a logarithmic measure of the acidity or alkalinity of a solution using the hydrogen ion concentration. The pH scale ranges from 0 to 14 with numbers above 7 being alkaline and numbers below
7. acidic. A one number change means the solution is ten times weaker or stronger than the previous measure.

- **phagocyte**

a type of white blood cell that can ingest invading microorganisms and other foreign material.

- **phase 1/11 trial**

these trials study the safety and dosage levels of a new treatment as in phase I trials, and the response of tumour to treatment as in phase II trials.

- **phase 11 trial**

these trials involve a larger number of participants than phase I trials and focus not only on side effects that may not have been seen in phase I trials, but also on tumour response to treatment.

- **phase 11/111 trial**
  these trials study tumour response to treatment as in phase II trials and the effectiveness of the treatment compared to the standard treatment regimen as in phase III trials.

- **phase i trial**
  in these trials, a small number of patients are given a new treatment to determine the safety of the treatment, including appropriate dosage levels. Because many of the treatments researched in phase I trials are new, participants may face substantial risks. Therefore, these trials are usually open only to patients in an advanced stage of cancer, whose disease is not controllable with standard treatment.

- **phase iii trial**
  once a treatment has been shown to be effective against a cancer, it is studied in a phase III trial to compare the new treatment with the standard treatment regimen. Safety, effectiveness, dosage, and side effects are all evaluated. More people are needed in phase III trials than in phase 11 trials to determine whether the new treatment is better than the standard treatment.

- **phase iv trial**
  once a treatment has been approved and is being marketed, it is studied in a phase IV trial to evaluate side effects of the new treatment that were not apparent in the phase III trial. A very large number of people are involved in a phase IV trial.

- **phenobarbital**
  a sedative/anticonvulsant barbiturate that has been used to treat diarrhea and to increase the anti-tumour effect of other therapies.

- **phenotype**
  observable characteristics resulting from interaction between an organism's genetic makeup and the environment. Compare Genotype.

- **phenylacetate**
  a drug that may suppress tumour growth.

- **phenylbutyrate**
  a drug used to prevent precancerous cells from developing into cancer.

- **phenylketonuria (PKU)**
abnormal presence of phenylketone and other metabolites of phenylalanine in the urine, characteristic of an inborn metabolic disorder caused by the absence or a deficiency of phenylalanine hydroxylase, the enzyme responsible for the conversion of the amino acid phenylalanine into tyrosine. Accumulation of phenylalanine is toxic to brain tissue. Untreated individuals have very fair hair, eczema, a mousy odor of the urine and skin, and progressive mental retardation. Treatment consists of a diet low in phenylalanine.

- **phosphate**
1. a normal chemical in blood and urine. Usually associated with stone disease as an inhibitor although in some stones such as calcium phosphate or struvite it can increase stone risk. One of the important chemicals to measure in any metabolic workup. Low levels of phosphate in the blood will cause an increase in Vitamin D activation and subsequent increase in intestinal absorption of phosphate and calcium.
2. an ion of phosphorus and oxygen (H2PO4- or HPO4-2). May exist as an ion in solution or as an ester or salt of phosphoric acid.

- **photodynamic therapy**
treatment that destroys cancer cells with lasers and drugs that become active when exposed to light.

- **photosensitiser**
a drug used in photodynamic therapy that is absorbed by tumour cells, when exposed to light, it becomes active and kills the cancer cells. Dihematoporphyrin ether, hematoporphyrin derivative, lutetium texaphyrin, and photofrin are photosensitising drugs.

- **photosynthesis**
1. conversion by plants of light energy into chemical energy, which is then used to support the plants' biological processes.

2. the reaction carried out by plants in which energy from sunlight is used to convert carbon dioxide from the atmosphere into sugars and oxygen.

- **physiologic**
related to the functions of the body. When used in the phrase 'physiologic age', it refers to an age assigned by general health, as opposed to the calendar age.

- **phytoremediation**
the use of plants to clean up pollution.

- **pica**
a craving for unnatural articles of food such as is observed in phosphorus-deficient animals, a depraved appetite.

- **pilot study**
the initial study examining a new method or treatment.

- **pink eye (conjunctivitis)**
an inflammatory disease of the eye of cattle of all ages. The eye may become pearl colored and opaque leading to blindness.

- **pipeline**
a stainless steel or glass pipe used for transporting milk by gravity to storage. May be above the milking units (high line) or below the level of the units (low line).

- **piritrexim**
an investigational chemotherapy drug.

- **pit**
a contained unit usually with concrete walls in which liquid or semi-liquid manure is stored.

- **placebo**
an inactive substance resembling a medication, given as a control in evaluating a medicine believed to be active. It is usually a tablet, capsule, injection, or infusion that contains a harmless substance but appears to be the same as the medicine being tested. A placebo may be compared with a new drug or treatment when no one knows if the new approach will be effective.

- **placenta**
the tissue attachment between the foetus and the mother.

- **plain**
a term suggesting general inferiority, coarse, lacking

the desired quality or breed character.

- **planning of activities**
  planning who is doing what and when.
- **plant capacity**
  means the maximum capability, on a short tons per day basis, of a processing or refining facility to process sugar beets (cleaned and tared), sugarcane and/or raw sugar.
- **plant community**
  a group of plants growing in a particular area under particular conditions of soil, climate etc. Communities can be classified in broad terms such as broadleaved woodland or in greater detail such as upland mixed ash woodland.
- **plasma**
  1. the fluid (non-cellular) fraction of blood.
  2. the clear, yellowish fluid portion of the blood in which cells are suspended.
  3. the liquid portion of blood or lymph in which corpuscles or blood cells float.
- **plasmapheresis**
  a technique used to separate useful factors from blood.
- **plasmid**
  a small circular form of DNA that carries certain genes and is capable of replicating independently in a host cell.
- **plasmin**
  plasmin is a proteolytic enzyme, it splits proteins. Plasmin attacks both ß -casein and alpha(s2)-casein. It is very heat stable and responsible for the development of bitterness in pasteurised milk and UHT processed milk. It may also play a role in the ripening and flavour development of certain cheeses, such as Swiss cheese.
- **plate cooler**
  a heat exchanger in which water at ground temperature or chilled water is used to cool milk prior to its movement to the bulk milk tank.
- **platelets**
  blood cells that help prevent bleeding by causing blood clots to form at the site of

an injury. An abnormally low number of platelets (called thrombocytopenia) may result in easy bruising and excessive bleeding from wounds or bleeding in mucous membranes and other tissues.

- **platinum**
  a metal that is an important component of some anti-cancer drugs, such as cisplatin and carboplatin.

- **pleura**
  two thin layers of tissue, one covering the lung and one lining the wall of the chest cavity. These two layers of tissue are separated by a small amount of fluid that acts as a lubricant, allowing the lungs to move smoothly during breathing.

- **pleural effusion**
  a collection of fluid between the thin layers of tissue lining the lung and the wall of the chest cavity (pleura). This fluid may contain cancer cells.

- **pleurodesis**
  treatment to eliminate the space between the tissues in the chest cavity and the membrane that lines the cavity to prevent the build-up of fluid in that space.

- **plot**
  a small piece of land.

- **plumage**
  feathers of a bird.

- **pluripotent cells**
  having the capacity to become any kind of cell or tissue in the body. Embryonic stem cells and cells of the inner cell mass are pluripotent. Adult stem cells are multipotent. The mammalian embryo (blastocyst trophoblast plus inner cell mass) is totipotent because it can become an entire organism. Fully differentiated cells from many plants are totipotent.

- **poaching**
  the trampling of land when wet, by stock, so the soil becomes churned and middy, often to the detriment of the vegetation. The illegal taking of GAME.

- **polled**
  describing a naturally hornless animal.

- **pollen**
  a fine powder produced by the 'male' parts of flowers

composed of spores that fertilise flowers of the same species.

- **pollen brush**
  a structure on the legs of some insects (most notably, bees) that is used to gather pollen.

- **pollution**
  the release of any harmful chemical into any of the environmental media (air, water, land).

- **polyclonal**
  derived from different types of cells.

- **polygastric**
  possessing more than one stomach compartment, as does the cow and other ruminants.

- **polymer**
  a long molecule of repeated subunits.

- **polymerase**
  general term for enzymes that carry out the synthesis of nucleic acids.

- **polymerase chain reaction (PCR)**
  1. a technique to amplify a target DNA sequence of nucleotides by several hundred thousandfold.
  2. a powerful technique for producing millions of copies of a specific region of DNA, so it can be analysed as readily as can a purified chemical substance. PCR has been instrumental in major breakthroughs in diagnostic kit development, forensic medicine, and detection of genes associated with inborn errors of metabolism. A Nobel Prize in medicine was awarded in 1993 for the development of PCR.

- **polyp**
  a growth on a mucous membrane.

- **polypeptide**
  long chain of amino acids joined by peptide bonds.

- **pondscape**
  a landscape containing a high density of ponds.

- **pondway**
  a string of ponds and associated landscape features that provide wildlife corridors for a range of aquatic and amphibious animals and plants.

- **pool, individual handler**
  a pool in which each producer selling raw milk to a single handler (dealer) receives the same blend price for milk regardless of the use made of any individual producer's milk.

- **pool, marketwide**
  a pool in which each producer supplying a market receives the same uniform price for milk calculated on the basis of use of all milk received by handlers in the market.

- **pooling**
  method used in determining how funds in a market will be distributed among producers supplying the milk.

- **pooling, cooperative**
  method of calculating prices received by members of a specific cooperative. If permitted by the bylaws, a cooperative may base the price it pays to members on the revenue obtained from sales fewer adjustments for operating expenses and reserves. As a result, the pooled price received by members of a cooperative may not necessarily correspond to federal or state marketing order announced pool prices. Government agencies do not interfere with the right of cooperatives to distribute returns to producers in accordance to membership contracts. In all other regards, cooperatives operating pool plants have the same obligations as proprietary handlers.

- **pooling, individual handler**
  method of pooling in which the price paid to producers is calculated for each handler based on how milk is used by the handler. In this type of pool, producers shipping to different handlers in the same market can receive different prices, depending upon the utilisation of the milk by the individual buyers.

- **pooling, marketwide**
  method of calculating the blend price paid to producers on the basis of the usage of all the milk received by all handlers in the market.

- **population**
  a group of organisms of the same species that live in the same area.

- **population density**
  the number of individuals in a particular population per unit area (e.g. per square km).
- **porfimer sodium**
  a drug used in some treatments to destroy cancer cells when exposed to light.
- **pork**
  the meat from hogs, or domestic swine. Much of a hog is cured and made into ham, bacon, and sausage. Uncured meat is called 'fresh pork'.
- **pork bellies**
  one of the major cuts of the hog carcass that, when cured, becomes bacon.
- **pork shoulder picnic**
  a front shoulder cut of pork which has been cured in the same manner as ham.
- **positive axillary lymph nodes**
  lymph nodes in the area of the armpit (axilla) to which cancer has spread. This is determined by surgically removing some of the lymph nodes and examining them under a microscope to see whether cancer cells are present.
- **post-legged**
  a condition in which the hind legs are too straight, so that the springy quality of the hock and pastern is lost.
- **post-milking teat dip**
  a product applied after milking to protect the teat from contagious pathogens that may have come into contact with the teat during the milking process.
- **postpartum**
  occurring after birth of the offspring.
- **potash**
  an oxide of potassium (K2O) used as a plant nutrient or fertiliser.
- **potentially hazardous foods**
  a food that is natural or synthetic and that requires temperature control because it is in a form capable of supporting the rapid and progressive growth of infectious or toxigenic microorganisms.
- **poult**
  a young turkey of either sex, usually not more than 2 months of age.

- **powder**
product obtained by removing water from pasteurised skim milk. It usually contains 3 to 4 percent moisture and 1 percent milk fat. Powder is also called skimmed milk powder in international markets. Powder is often abbreviated 'NDM' or 'NFDM', which stands for non-fat dry milk.

- **ppb**
parts per billion. It equals micrograms per kg or micro liters per liter.

- **ppm**
parts per million. It equals milligrams per kg or milliliters per liter.

- **preclinical studies**
studies that test a drug on animals and in other non-human test systems. Safety information from such studies is used to support an investigational new drug application (IND).

- **predation**
the interaction between populations in which one organism (the predator) consumes another (the prey). Typically, the predator catches, kills and eats its prey but predation is also used to describe feeding by insectivorous plants and even grazing by herbivores.

- **predicted producing ability (PPA)**
an estimation of the amount of milk and/or components that a cow will yield above or below herd mates based on the cow's pedigree information and performance, if available.

- **predicted transmitting ability (PTA)**
an estimation of an animal's potential to transmit yield productivity to its offspring based on pedigree information and the animal's performance, if available.

- **predisposition**
the state of being particularly susceptible to a certain condition or disease.

- **prednisolone**
a synthetic corticosteroid used in the treatment of blood cell cancers (leukemias) and lymph gland cancers (lymphomas).

- **prednisone**
belongs to the group of drugs known as steroids. It

is used to treat several types of cancer. Prednisone also inhibits the body's immune response and is not allowed as treatment for patients participating in some clinical trials.

- **preliminary milking**
  the first milking of a three-milking verification test for herds milked twice daily.

- **premalignant**
  precancerous, changes have occurred that may develop into cancer.

- **premature calving**
  termination of pregnancy during the last trimester.

- **premier sires**
  livestock Improvement's team of elite bulls.

- **pre-milking teat dip**
  a product applied in preparation for milking to clean the teat and reduce the spread of disease and maintain healthy teats.

- **prepared**
  slaughtered, canned, salted, rendered, boned, cut up, or otherwise manufactured or processed.

- **prepartum**
  occurring before birth of the offspring.

- **prepotent**
  designating an animal that transmits its own character to its progeny to a marked or highly uniform degree.

- **prescription**
  written instructions. For example, prescriptions in a nature reserve management plan state how it is intended the reserve or a compartment of the reserve should be managed.

- **preservation**
  a variety of methods used at the processing stage and at home to keep food safe from harmful bacteria and extend the storage life of food.

- **pricing formula**
  mathematical equations that use economic data to establish a minimum regulated price.

- **primary central nervous system lymphoma**
  cancer of the lymph tissue in the central nervous system (brain, spinal cord and meninges).

- **primary forest**
  the surviving fragments of primaeval forests, the climax vegetation type of this country.

- **primary tumour**
  the tumour is located in the place where the cancer first started to grow.

- **primary waste treatment**
  usually, the initial treatment of waste water to reduce pollution potential or health hazard. With animal manure an example would be liquid-solid separation.

- **primitive neuroectodermal tumours**
  tumours that originate in bone or soft tissue.

- **principles of a group**
  features of a participatory group (by ICA).

- **probiotics**
  live friendly bacterial cultures, added to foods such as yogurt, for health benefits including assisting digestion and positive gut health. 'Pro' = positive/for + 'biotic' = life. The probiotics most commonly added to yogurt are bifidobacteria and lactobacilli.

- **procarbazine**
  an anticancer drug used in combination chemotherapy.

- **process cheese**
  a blend of natural cheeses melted together and heated to prevent further ripening. This process gives process cheese uniform flavour, body and texture.

- **process cheese food**
  prepared like process cheese, except it contains slightly more moisture and less milk fat than process, cheese. Other dairy products such as milk, cream, and buttermilk may be added to the process cheese. This product must be at least 51% natural cheese by weight,

- **process cheese spread**
  prepared like process cheese food except it is soft and spreadable and contains less fat than process cheese food.

- **processing facility**
  means a distinct physical facility, at a single location, which processes sugarcane, sugar beets or molasses into sugar.

- **processing inputs**
  means the quantity of raw materials (e.g., sugarcane,

sugar beets, raw sugar, and molasses) used in processing or refining operations.

- **processing of milk**
making milk products like cheese, yoghurt etc.

- **produce a female's offspring.**
the produce of dam commonly refers to two offspring of one dam.

- **producer**
any person that produces milk from five or more cows in conformity with the applicable health regulations of the place in which it is sold, and whose bulk market milk is received, acquired, or handled by any handler or any non-profit association of producers.

- **product**
any carcass, meat, meat byproduct, or meat food product, capable of use as human food.

- **production**
means the output of beet sugar from the processing by sugar beet processors of domestically produced sugar beets or sugar beet molasses. The output of cane sugar (including edible molasses and cane syrup) by sugarcane processors of domestically produced sugarcane or sugarcane molasses, or the output of sugar (including edible molasses and sugar syrup) from the processing by cane sugar refiners of raw cane sugar or imported molasses.

- **production and type index (pti)**
combines genetic merit measurements for production and type in a ratio thus ranking sires on their ability to transmit a balance of these traits.

- **productivity**
ability to produce (in the case of a cow) milk.

- **progeny tested**
a programme which generates daughters of promising young bulls that are milked and tested for milk production and a number of other traits.

- **progeny testing**
evaluating the genotype of an individual by a study of its offspring. An evaluation of the transmitting ability of an individual based on the performance of offspring.

- **prognosis**
  the likely outcome or course of a disease, the chance of recovery.

- **progressive disease**
  cancer that is increasing in scope or severity.

- **projected 305-day lactation**
  a calculation for predicting a cow's total yield in 305 days based on the information from a lactation in progress.

- **prokaryote**
  an organism (e.g., bacterium, virus, cyrobacterium) whose DNA is not enclosed within a nuclear membrane. Compare eukaryote.

- **prolapsed uterus**
  a condition in which the uterus is partially or completely turned inside out, outside the animal's body, usually following parturition.

- **promoter**
  a DNA sequence that is located in front of a gene and controls gene expression. Promoters are required for binding of RNA polymerase to initiate transcription.

- **propagate**
  to increase the number of plants of a particular population by means of cuttings.

- **proprietary handler**
  a milk processing company owned privately or publicly by investors other than milk producers. The business differs from a cooperative.

- **prosciutto ham**
  an Italian-style dry cured raw ham, not smoked, often coated with pepper. Prosciutto can be eaten raw because of the way it is processed.

- **protease**
  an enzyme that digests proteins.

- **protease inhibitor**
  1. a drugs that interferes with the ability of a virus to make copies of itself. Protease inhibitors include indinavir sulfate, ritonavir, saquinavir mesylate and nelfinavir mesylate.
  2. a new group of medications used in the treatment of AIDS. Unfortunately, some tend to form stones.

▪ protein

1. a molecule composed of amino acids. There are many types of proteins, all carrying out a number of different functions essential for cell growth.

2. one type of compound found within the body. Proteins make up much of the body's tissue in addition to being the main part of enzymes, hormones, and immunologic substances

3. a linear polymer that, in its biologically active state, functions as a catalyst in metabolism or as a structural element of cells and tissues.

▪ protein a

a protein produced by the bacterium Staphylococcus aureus that specifically binds antibodies. It is useful in the purification of monoclonal antibodies.

▪ protein engineering (genetic engineering)

the alteration of DNA to encode the synthesis of new proteins that perform new functions. DNA is made of only four nucleotides, in varying sequences, so it can be modified more easily than can proteins, which are difficult to alter using standard chemical means. Protein engineering was the basis for a Nobel prize in 1994.

▪ protein equivalent

a term indicating the total nitrogen content of a substance in comparison with the nitrogen content of protein (usually plant). For example, the nonprotein nitrogen (NPN) compound, urea, contains approximately 45 percent nitrogen and has a protein equivalent of 281 percent (6.25 x 45 percent).

▪ protein pricing

one variation of a component (or multiple component) pricing system that involves establishing a price for the protein portion of milk. A protein pricing formula will often assign a portion of the total value to the fat portion and to the other non–protein solids as well.

▪ protein supplements

feed products that contain 20 percent or more of protein.

▪ proteomics

each cell produces thousands of proteins, each with

a specific function. This collection of proteins in a cell is known as the proteome, and, unlike the genome, which is constant irrespective of cell type, the proteome varies from one cell type to the next. The science of proteomics attempts to identify the protein profile of each cell type, assess protein differences between healthy and diseased cells and uncover not only a protein's specific function but also how it interacts with other proteins.

- **protocol**
  a set of rules governing how something is to be done.

- **protoplast**
  the cellular material that remains after the cell wall has been removed from plant and fungal cells.

- **proven sire**
  a bull whose genetic transmitting ability has been measured by comparing the milk production performance of his daughters with that of the daughter's dam and/or herd mates under similar conditions. Plus-proven sires are those whose daughters exceeded their dams and/or herd mates in milk production.

- **proximate analysis**
  tests for nitrogen (crude protein), crude fibre, ether extract (lipids), and ash which, with nitrogen-free extract, represent the gross composition of feed.

- **psc 833**
  belongs to a group of drugs known as cyclosporine analogues. Used with chemotherapy to prevent or overcome the resistance of tumour cells to some chemotherapy drugs.

- **psychrophilic cold-loving.**
  it refers to microorganisms that grow best at refrigerated temperatures, that is, 4 to 15°C.

- **psychrotrophic**
  said of an organism capable of growing at temperatures close to freezing, or having an optimum temperature that is low.

- **psychrotrophic cold-tolerant.**
  it refers to microorganisms that grow at low temperatures, 4 to 15°C., but may have an optimum tempera-

ture above this range. These organisms especially affect the shelf life of refrigerated dairy products such as cottage cheese.

- **PTA**

Predicted Transmitting Ability is the predicted difference of a parent animal's offspring from average, due to the genes transmitted from that parent. Each PTA is given in the units used to measure the trait. The PTA for milk is reported in pounds or kg, the PTA for productive life is reported in months.

- **public health**

the science and the art of 1) preventing disease 2) prolonging life, and organised community efforts for a) the sanitation of the environment b) the control of communicable infections c) the education of the individual in personal hygiene d) the organisation of medical and nursing devices for the early diagnosis and preventive treatment of disease and e) the development of the social machinery to ensure everyone a standard of living adequate for the maintenance of health, so organising these benefits as to enable every citizen to realise his/her birthright of health and longevity.

- **pullet**

a female chicken less than 1 year old, or a young female chicken before she begins to lay eggs.

- **pulmonary**

refers to the lungs.

- **pulse-field gel electrophorsis (PFGE)**

the DNA fingerprinting method that scientists use to determine the source of bacteria in foods.

- **pupa**

the stage in an insect lifecycle between larva and adult. This is the stage during which metamorphosis takes place and pupae are often covered in a hard covering and are apparently inactive. Under the covering a radical re-arrangement of body structures occurs to alter, for example, the caterpillar into the adult butterfly or moth. Some insects overwinter as pupae.

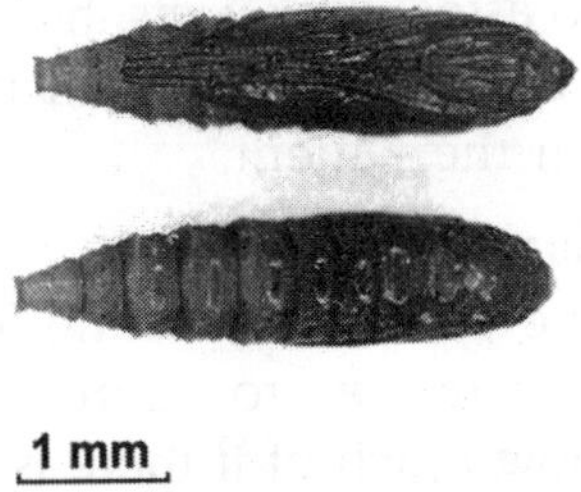

■ **pupal**
relating to the pupa.

■ **pupate**
to go through the pupal stage in an insect lifecycle.

■ **pure culture**
in vitro growth of only one type of microorganism.

■ **purebred**
the offspring of purebred parents of the same breed. Ancestors of purebreds can be traced to foundation stock in the original herd book. An animal with two registered parents of the same breed. Purebreds are not all registered.

■ **purebred dairy cattle association (PDCA)**
an organisation made up of members of the six purebred dairy cattle registry associations: Ayrshire, Brown Swiss, Guernsey, Holstein, Jersey and Milking Shorthorn.

■ **purine**
a component of the genetic code material (DNA) inside every cell. All proteins have some purine, but the highest levels are found in red meat, poultry and fish. Digestion of excessive amounts of high purine content proteins can create high levels of uric acid.

■ **pus**
a liquid product of inflammation consisting of leukocytes, lymph, bacteria, deadtissue cells and fluid derived from their decomposition.

■ **pyelonephritis**
an infection of the kidney. Can progress to sepsis or bacterial blood poisoning.

■ **pyometra**
an accumulation of pus in the uterus.

■ **pyrazine diazohydroxide**
an investigational chemotherapy drug.

■ **pyrazoloacridine**
belongs to a group of anticancer drugs known as acridines.

■ **qs21**
a plant extract that can improve the ability of the im-

mune system to respond to disease, often used with vaccine therapy.

- **qualitative analysis**
  the process of testing for a substance to determine what it is and what its components are. The results are reported in terms of the presence or absence of particular components, based on the size of the sample used in the analysis, the number of samples analysed, and the testing method. An example of qualitative analysis would be testing for the presence of the bacterial pathogen Listeria monocytogenes in a specific food.

- **qualitative risk assessment**
  a risk assessment that is based on qualitative data or giving a qualitative result. The results are often stated in an estimated range, such as 'there is a moderate to high risk of a certain outcome occurring'.

- **quality of life**
  the overall enjoyment of life. Many clinical trials measure aspects of a patient's sense of well-being and ability to perform various tasks to assess the effects that cancer and its treatment have on the patient.

- **quantitative analysis**
  the process of testing for a substance to determine how much of it there is and the numerical value of each of its components. An example would be testing for the amount or concentration of a certain chemical or microorganism, such as E. coli, in a food.

- **quantitative risk assessment**
  a risk assessment that uses modeling to determine the probability(s) of what can go wrong, how likely it is to happen, and how severe is the health impact. The results are stated in numerical terms, such as 'there is a 42% probability that one illness may occur from eating a serving of X food with a certain health outcome'.

- **quantitative trait loci**
  a combination of genes that often controls economically significant genetic traits, such as disease resistance in animals and, in dairy cattle, milk quality and quantity.

- **quarg or quark**
smooth curd, unripened fresh cheese made from skim milk. Used to make fromage frais for Petit Miam and Frûche.

- **quinoa**
a goosefoot (Chenopodium quinoa) native to the Andes and cultivated for its edible seeds. The high-protein dried fruits and seeds of this plant, used as a food staple and ground into flour.

- **race**
a track between paddocks.

- **radiation surgery**
a radiation therapy technique that delivers a single high dose of radiation directly to the tumour while sparing the healthy tissue. Also called radio surgery and stereotactic external beam irradiation.

- **radiation therapy**
radiation therapy (also called radiotherapy) uses high-energy radiation from x-rays and other sources to kill cancer cells and shrink tumours. Radiation may come from a machine outside the body (external radiation therapy) or from materials that produce radiation (radioisotopes) that are placed inside the body in the area where the cancer cells are found (internal radiation therapy, implant radiation, or brachytherapy). Systemic radiation therapy involves giving a radioactive substance, e.g., a radio labelled monoclonal antibody that circulates throughout the entire body.

- **radioactive iodine**
a radioactive form of the chemical element iodine often used for imaging tests or as a treatment for cancer.

- **radioimmunoassay (RIA)**
a test combining radioisotopes and immunology to detect trace substances. Such tests are useful for studying antibody interactions with cell receptors, and can be developed into clinical diagnostics.

- **radioimmunotherapy**
treatment with a radioactive substance that is linked to an antibody that will attach to the tumour when injected into the body.

- **radioisotope**
  an unstable molecule that releases radiation as it decays. Can be used in testing or as treatment.

- **radiolabelled**
  an antibody that has been joined with a radioactive substance (often iodine).

- **radiosensitisation**
  the use of a drug to make cells more sensitive to radiation therapy.

- **raised bog**
  a type of BOG which in cross-section is a shallow dome-shape. As a result, the bog surface (at least in the center) is raised above the normal level of the groundwater.

- **raltitrexed**
  an investigational chemotherapy drug that inhibits tumour cells from multiplying by interfering with cells' ability to make DNA. Also called ICI DI 694 and Tomudex.

- **ram**
  a male sheep that has not been castrated, generally employed for breeding purposes.

- **ramsar site**
  wetlands of international importance designated under the Ramsar convention 1971, which requires signatory countries to protect internationally important wetlands, especially those used by migratory water birds and to use wetlands wisely.

- **rancid/rancidity**
  oxidation/breakdown of fat that occurs naturally, causing undesirable smell and taste.

- **randomised clinical trial**
  a study in which participants are assigned by chance to separate groups that compare different treatments. Using chance to assign people means that the groups will be similar and the treatments they receive can be compared. At the time of the trial, there is no way for the researchers to know which of the treatments is best. It is the patient's choice to be in a randomised trial or not

- **range**
  the area within which a species or sub-species is found.

■ **ranking**
method used to prioritise items.

■ **rapid exit**
panels or rails that raise to release all cows on one side of the milking parlour at once.

■ **raptor**
a bird of prey.

■ **ras gene**
a gene that has been found to cause cancer when it is altered (mutated). Agents that block its activity may stop the growth of cancer. A ras peptide is a protein fragment produced by the ras gene.

■ **rate of passage**
the time taken by undigested residues from a given meal to reach the faeces.

■ **ration**
the food allowed an animal for 24 hours.

■ **rational drug design**
using the known three-dimensional structure of a molecule, usually a protein, to design a drug molecule that will bind to it. Usually viewed as an alternative to drug discovery through screening many molecules for biological activity.

■ **raw milk**
farm milk that has not been treated in any way. Raw milk is not pasteurised, separated, standardised or homogenised.

■ **raw milk**
fresh, unprocessed milk as it comes from a cow, goat or other mammal.

■ **raw sugar**
means any sugar, whether or not principally of crystalline structure, which is to be further refined or improved in quality.

■ **ready-to-eat**
food that is in a form that is edible without washing, cooking, or additional preparation by the food establishment or consumer and that is reasonably expected to be consumed in that form.

■ **reagent**
substance used in a chemical reaction.

■ **rebeecamycin**
a drug that interferes with tumour cell DNA and inhibits tumour growth.

recalcitrant
in plant nutrition, refers to that portion of a nutrient in the soil that is unavailable for biochemical activity or plant uptake.

receipts
refers to the quantity of raw materials (e.g., sugarcane, sugar beets, raw sugar, refined sugar, liquid sugar, syrups, and molasses) received by the processing or refining facility.

receiving station
facility at which milk is collected from farm bulk milk trucks, stored, then shipped, usually in tractor–trailer units, to another destination.

receptive
willing to receive, accepting.

receptor site
the site to which individuals of a species are taken in order to be introduced from another place (the 'donor site').

reclaimed
land that has been brought into new productive use. It is commonly applied to land previously in the intertidal zone that has been brought into agricultural use by a combination of drainage and installation of sea defences.

recombinant dna
the hybrid DNA produced by joining pieces of DNA from different organisms together in vitro. This is accomplished by inserting a desired DNA sequence into a vector (usually a circular strand of DNA or a virus), which then is introduced into another organism, often a bacterium.

recombinant DNA (rDNA)
the DNA formed by combining segments of DNA from two different sources.

recombinant DNA technology
a broad term referring to molecular cloning as well as techniques for making recombinant DNA or using it for specific purposes.

reconstituted milk
product resulting from the mixing together and rehydration of a dried product of milk with water. For example, non-fat dry milk and water

yields reconstituted skim milk. Adding cream or butter oil yield reconstituted whole or low fat milk. Adding vegetable oil yields filled milk.

- **reconstituted milks**
  the product resulting from the mixing together and rehydration of a dried product of milk with water. For example, NDM and water yield reconstituted skim milk.

- **record**
  a biological record is a document or other thing (e.g. entry on computerised database) that preserves information about some biological entity.

- **record keeping**
  noting down important information for the group.

- **recurrence**
  the return of cancer after it had apparently completely disappeared.

- **recurrent cancer**
  cancer that has come back (recurred) after it has been treated. It may come back in the site of the original cancer or in another part of the body.

- **red data book**
  contains details about nationally rare species.

- **red list**
  the list of species covered within a red data book.

- **red meat**
  refers to the carcass after slaughter for beef, veal, pork, lamb and mutton.

- **reduced fat**
  a product that contains at least a 25% reduction in total fat per reference amount when compared to an appropriate reference food.

- **reedbed**
  wetland dominated by stands of common reed (Phragmites australis).

- **refractory cancer**
  cancer that does not respond to treatment.

- **regeneration**
  new growth. Laboratory technique for forming a new plant from a clump of plant cells.

- **regimen**
  a treatment plan that specifies the dosage, the schedule, and the duration of treatment.

- **regional cancer**
  refers to cancer that has grown beyond the organ of origin to regional lymph nodes and/or organs and tissues next to the original site.

- **regional quota adjusters**
  represent deductions that apply to quota payments to producers. They are determined by the location of the producer's dairy and apply to the hundredweight equivalent of quota milk produced.

- **regionally important geological/geomorphological site (RIGS)**
  sites, excluding SSSIs, that are considered worthy of protection for their educational, research, historical or aesthetic importance because of their geology.

- **registered**
  purebred animal registered in the herd book of the proper breed association. Certain associations (for example, the Red and White association) will register non-purebred cattle providing they satisfy certain other criteria.

- **regulatory gene**
  a gene that acts to control the protein-synthesising activity of other genes.

- **regurgitate**
  to cast up undigested food from the stomach to the mouth, as done by ruminants.

- **relapse**
  cancer that has come back after it has been treated.

- **reliability**
  sometimes called repeatability.

- **reliability (rel)**
  the term used in dairy cattle breeding to indicate the precision with which genetic merit is estimated. Reliability, then, is a confidence factor which increases with the (1) number of daughters, (2) number of herds in which daughters are located, (3) number of lactations per daughter, and (4) information of relatives used.

- **remission**
  disappearance of the signs and symptoms of cancer. When this happens, the disease is said to be 'in remis-

sion'. A remission may be temporary or permanent

- **renal**
anything to do with the kidney.

- **renal failure**
the loss of the kidneys ability to adequately filter the blood, usually due to diabetes or other diseases. When severe, requires mechanical filtration (dialysis) or kidney transplant.

- **renal leak hypercalciuria**
a type of increased urinary calcium due to a defect in the kidney that causes excessive calcium loss. Parathyroid hormone levels will tend to be elevated. This type of high urinary calcium doesn't respond to dietary modifications and requires medical therapy. Thiazides are particularly useful here. It is important never to use Cellulose therapy with renal leak hypercalciuria because it will cause a severe calcium loss.

- **renal phosphate leak hypercalciuria**
condition where the kidney is unable to keep excessive amounts of phosphate from being lost in the urine. This causes a low serum phosphate, high Vitamin D levels and ultimately high urinary calcium (hypercalciuria).

- **renal tubular acidosis**
condition in which the kidneys are unable to excrete acid. It's characterised by an inability to acidify the urine even after an oral acid load. Associated with extremely low urinary citrate and severe stone disease.

- **renderer**
a business engaged in the separation of fats from animal tissue by heating.

- **rennet**
  1. substance causing milk to coagulate, used for the preparation of cheese.
  2. enzyme used to precipitate the casein protein in milk to make cheese.
  3. an extract of the stomach of certain mammals containing rennin. It is used in the coagulation of milk proteins in the manufacture of most cheeses.

- **replacements heifers**
often ones being raised to replace the cows currently in the herd.

- **replication**
reproduction or duplication, as of an exact copy of a strand of DNA.

- **replicon**
a segment of DNA (e.g., chromosome or plasmid) that can replicate independently.

- **repressor**
a protein that binds to an operator adjacent to a structural gene, inhibiting transcription of that gene.

- **resected**
surgically removed.

- **residual disease**
cancer cells that remain after attempts have been made to remove the cancer.

- **residue**
any substance, including metabolites, remaining in livestock at time of slaughter or in carcass tissues after slaughter as the result of treatment or exposure of the livestock to a pesticide, organic or inorganic compound, hormone, hormone like substance, growth promoter, antibiotic, anthelmintic, tranquiliser, or other therapeutic or prophylactic agent.

- **resorptive hypercalciuria**
a condition of high urinary calcium where the source of the calcium is reabsorption from existing bodily stores such as bone. Most commonly caused by hyperparathyroidism.

- **restricted use market milk**
(degraded milk) Any milk produced at a market milk dairy, which does not conform to or was not produced under the standards established for market milk.

- **restriction enzyme**
an enzyme that breaks DNA in highly specific locations, creating gaps into which new genes can be inserted.

- **restriction fragment length polymorphism (RFLP)**
the variation in the length of DNA fragments produced by a restriction endonuclease that cuts at a polymorphic locus. This is a key tool in DNA fingerprinting and is based on the presence of different alleles in an in-

dividual. RFLP mapping is also used in plant breeding to see if a key trait such as disease resistance is inherited.

- **retail**
  final link in the marketing chain, refers to the interface between consumers and vendors of products packaged in their final form.

- **retained placenta**
  placental membranes not expelled normally at parturition.

- **reticuloendothelial system**
  the system of macrophages, which serves as an important defence system against disease.

- **reticulum**
  the second division or stomach compartment of a ruminant. Also called a honeycomb.

- **retinoid**
  vitamin A or a vitarnin A-like compound.

- **retrograde pyelogram**
  a special type of X-ray in which a small slender tube is placed through the bladder into the lower end of the ureter and contrast injected directly while an X-ray picture is taken. Alough this procedure requires anesthesia or sedation, the technique gives a very clear picture of the ureters and kidney interior. Can be used safely even when patients are severely allergic to contrast because there is no contact with the bloodstream.

- **retroviral vector**
  rNA from a virus that is used to insert genetic material into cells.

- **retrovirus**
  a virus that contains the enzyme reverse transcriptase. This enzyme converts the viral RNA into DNA, which can combine with the DNA of the host cell and produce more viral particles.

- **return alley**
  the alley through which cows must pass when moving from the milking parlour back to the cow housing area after milking.

- **reynold's number**
  a dimensionless expression used in predicting flow patterns.

- **rheology**
  study of the flow of matter such as fermentation liquids.

- **rhizobium**
  a class of microorganisms that converts atmospheric nitrogen into a form that plants can utilise for growth. Species of this microorganism grow symbiotically on the roots of certain legumes, such as peas, beans and alfalfa.

- **rhizome**
  an elongated underground stem in some plants that allows them to overwinter and to spread throughout a habitat (e.g. bracken, mint).

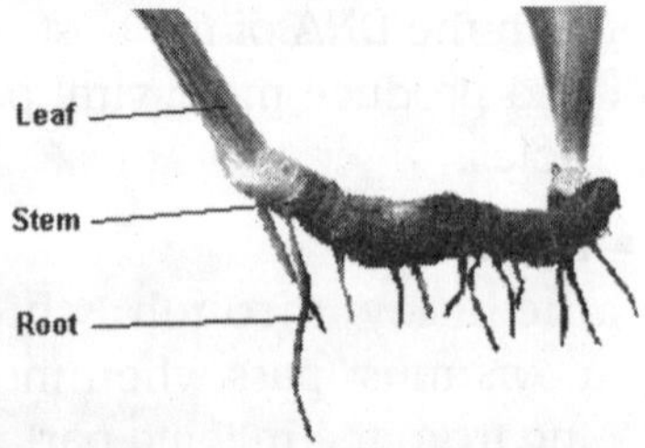

- **ribonucleic acid (RNA)**
  a molecule similar to DNA that delivers DNA's genetic message to the cytoplasm of a cell where proteins are made.

- **ribosome**
  a cellular component, containing protein and RNA, that is involved in protein synthesis.

- **ride**
  1. cleared area of woodland, often linear, for access, fire breaks and to provide open areas for game and wildlife.
  2. to be transported in a mounted position, as one cow mounted on another during estrus.

- **riding the pool**
  techniques by which a handler whose major concern is the production of manufactured products can arrange his business so as to legally participate in the pool. A handler (or a cooperative) 'rides the pool' by disposing of just enough of his milk supply to Class 1 plants to qualify as a pool plant. In this way, the handler is able to return the generally higher pool price to producers than if the handler simply engaged in a routine manufacturing business. It has advantages because the handler can use the pool to

pay more for the milk than what the manufacturing class prices would indicate.

- **riffle**
  a bar of material with a rippled surface deposited on the bed of a river or stream. The smooth flow of water is disrupted when it passes over a riffle causing it to mix and allowing it to take up more oxygen. Riffles can harbour rich assemblages of freshwater invertebrates and mosses.

- **ring feeder**
  a steel hoop with individual head gates that may be placed over a large round bale of hay when feeding it.

- **ring test**
  a test for brucellosis performed by mixing stained Brucella bacteria with raw milk. If antibody to Brucella is present, the stained cells agglutinate (clump) and rise to the surface with the cream to form a blue ring.

- **riparian**
  relating to or situated on the bank of a river or stream.

- **risk analysis**
  the assessment and management of hazards that cause harm (risk) to human health and the communication of how those hazards can be controlled, reduced or eliminated.

- **risk assessment**
  the process of estimating the severity and likelihood of harm to human health or the environment occurring from exposure to a substance or activity that, under plausible circumstances, can cause harm to human health or the environment.

- **risk communication**
  exchanges of information among risk assessors, risk managers, other stakeholders, and the public about levels of health or environmental risk, the significance and meaning of those risks, and the decisions, actions, or policies aimed at managing or controlling the risks.

- **risk factor**
  anything that increases a person's chance of developing a disease.

- **risk management**
  the process of evaluating policy alternatives in view of the results of risk assessment and selecting and

implementing appropriate options to protect public health. Risk management determines what action to take to reduce, eliminates, or control risks. This includes establishing risk assessment policies, regulations, procedures, and a framework for decision-making based on risk.

- **rmp-7**
  a drug that allows chemotherapy drugs (such as carboplatin) to reach tumours in the brain.

- **RNA**
  ribonucleic acid, which is similar to a single strand of DNA but with a small chemical difference in the structure of the nucleotides. There are three classes: messenger RNA is the template for decoding the instructions contained in DNA, while transfer RNA and ribosomal RNA (located in structures known as ribosomes inside cells) follow these instructions to make proteins.

- **RNA interference**
  a natural process used by organisms to block protein production.

- **RO (reverse osmosis)**
  a process whereby a liquid such as milk or whey is pumped through a semi-permeable membrane under high pressure. The size of the pores in the membrane allows separation to occur at the molecular level. The liquid that passes through the membrane is called permeate, and the remainder is called retentate. With RO, the permeate is primarily pure water. Commonly used to reduce the volume of milk or whey prior to drying or further processing.

- **roan**
  a close mixture of red and white colors such as is often found in Milking Shorthorn cattle.

- **roaster duckling**
  a young duck, usually under 16 weeks of age, of either sex, that is tender-meated and has a bill that is not completely hardened and a windpipe that is easily dented.

- **roaster or roasting chicken**
  a bird of this class is a young chicken, usually three to

five months of age, of either sex, that is tender-meated with soft, pliable, smooth-textured skin and breast-bone cartilage that may be somewhat less flexible than that of a broiler or fryer.

- **robotic parlour**
  a completely automated system for milking cows that requires limited human contact.

- **rock (cornish game hen)**
  a Rock Cornish game hen or Cornish game hen is a young immature chicken (usually five to six weeks of age) weighing not more than two pounds ready-to-cook weight, which was prepared from a Cornish chicken or the progeny of a Cornish chicken crossed with another breed of chicken.

- **rock (cornish fryer, roaster, or hen)**
  a Rock Cornish fryer, roaster, or hen is the progeny of a cross between a purebred Cornish and a purebred Rock chicken, without regard to the weight of the carcass involved; however, the term 'fryer', 'roaster', or 'hen' shall apply only if the carcasses are from birds with ages and characteristics of a 'broiler or fryer' or 'roaster or roasting chicken'.

- **role of ca++**
  more than 90% of the calcium content of skim milk is associated in some way or another with the casein micelle. The removal of Ca++ leads to reversible dissociation of ß -casein without micellular disintegration. The addition of Ca++ leads to aggregation.

- **rolling herd average (RHA)**
  the average milk production per herd per year based on the 12 months just past. Upon completion of a new test record, the record for the same period of the previous year is deducted and the new record is added, then a new rolling 365-day average is calculated. RHA's are updated with each new test.

- **roost**
  a place where birds or bats rest or sleep (often in groups).

- **rooster**
  an adult male chicken.

■ **rootstock**
a rhizome.

■ **rotary milking parlour**
a raised, round rotating platform or carousel on which cows ride while being milked.

■ **rotary parlour**
a raised, round rotating platform or carousel on which cows ride while being milked.

■ **rotation diet**
a diet in which food ingredients are eaten only once every few days. This spread out approach allows for more variation in the diet and decreases potential allergy or intolerance reactions since the substance is not allowed to build up in the body.

■ **rotational set-aside**
a management regime whereby the location of set-aside land is periodically switched from place to place within a farm holding.

■ **roughage**
consists of pasture, silage, hay, or other dry fodder. It may be of high or low quality. Roughages are usually high in crude fiber (more than 18 percent) and relatively low in NFE (approximately 40 percent).

■ **rpr 109881a**
belongs to a group of anticancer drugs called taxanes.

■ **RSR**
a drug that increases the effectiveness of radiation therapy. Also called RSR- 1 3.

■ **RSR-13**
a drug that increases the effectiveness of radiation therapy.

■ **rugged**
when referring to an animal, it means large and strong.

■ **rumen**
the first and largest stomach compartment of a ruminant, also called paunch.

■ **ruminant**
1. an animal with a stomach that has four compartments, and a more complex digestive system than other mammals. Ruminants include cattle, sheep, goats, deer, bison, elk, and camels. Swine, dogs and humans

are examples of non-ruminants.

2. animals having four stomach compartments-rumen, reticulum, omasum and abomasums-through which food passes in digestion. These animals chew their cud or regurgitate partially digested food for further breakdown in the mouth. Ruminant animals include cattle, sheep, goats, deer and camels.
3. one of the order of animals having a stomach with four complete cavities—rumen, reticulum, omasum, and abomasum—through which food passes in digestion. These animals chew their cud. They include cattle, sheep, goats, deer and camels.

- **rumination**
  the casting up or regurgitation of food (cud) to be chewed a second time, as cattle do. A chewing of the cud by ruminants.

- **run on**
  to graze or pasture on, as for cattle to run on the range.

- **runnel**
  a groove in a surface that carries water.

- **rush**
  a member of the plant family Juncaceae.

- **rushy**
  dominated by rushes.

- **rye-grass mixture**
  an agricultural seed mix containing fast-growing Rye Grass (Lolium perenne).

- **sacrificial seed crop**
  seed that would normally be sown for a commercial harvest (e.g. cereal) that is sown instead with the intention of leaving it to be consumed by animals such as see-eating birds.

- **salmonella**
  1. any of various bacteria causing food poisoning, typhoid, and other infectious diseases.
  2. a pathogenic, diarrhoea-producing bacterium that is the leading cause of human foodborne illness among intestinal pathogens. It is commonly found in raw meats, poultry, milk, and

eggs, but other foods can carry it.

- **salmonid**
  a member of a large family of freshwater fish species that includes trout and salmon.
- **salvage chemotherapy**
  chemotherapy given after the primary treatment has failed to eliminate all of the cancer or when the cancer returns after having been in remission.
- **samarium 153**
  a radioactive substance used in cancer therapy.
- **sample**
  a specimen that is taken from food and tested for the purpose of identifying a foodborne pathogen or various kinds of chemical contaminants in food.
- **sand separator**
  a mechanical device or series of course ways used to settle sand from sand-laden manure.
- **sanitise**
  to kill or remove injurious, microorganisms but not necessarily to sterilise. Dairy equipment is commonly sanitised with heat or chemicals.
- **sanitiser**
  chemical or physical agents that reduce microorganism contamination levels present on inanimate environmental surfaces.
- **sapling**
  a young tree.
- **sarcoma**
  a cancer of the bone, cartilage, fat, muscle, blood vessels or other connective tissue.
- **saving**
  putting money aside for future purposes.
- **SBR grade**
  a histologic grading system by which tumour cells are grouped based on their appearance when compared with normal cells.
- **scale**
  the overall size of an animal.
- **scale-up**
  transition from small-scale production to production of large industrial quantities.
- **sch-58500**
  an investigational drug that inhibits the growth of tu-

mour cells that express the mutated p53 gene.

- **scientific name**

the unique two-part latin name given to each species that has been described at some time in a scientific journal. For example, the water vole's scientific name is 'Arvicola terrestris'. The 'Arvicola' part denotes that the vole is classified as belonging to the genus 'Arvicola'. The second part of the name 'terrestris' allows this animal to be distinguished from other species in the same genus. The Scientific Name is useful to allow naturalists from different parts of the world to be sure they are talking about the same animal or plant. For example, the 'water vole' is known as the 'water rat' to some people in this country and, even in identification guides written in English, has been called variously 'water vole', 'northern water vole' and even 'ground vole'. Some authors have used the term 'water vole' when talking about a completely different but closely related species (Arvicola sapidus). Consistent use of the Scientific Name is meant to minimise confusion.

- **scours**

a persistent diarrhoea in animals.

- **scrambling**

the form of growth of certain plants (e.g. purple ramping-fumitory) that have long weak shoots and grow over other plants using them for support.

- **scrapie**

a fatal, degenerative neurological disease of sheep and goats. Belonging to a family of diseases known as transmissible spongiform encephalopathies (TSEs), scrapie is similar to bovine spongiform encephalopathy (BSE) (mad cow), a disease of cattle. There is no scientific evidence to indicate that scrapie poses a risk to human health.

- **scrub**

an animal from nonpurebred parents not showing the predominant characteristics of any breed.

- **scrub-over**
grassland that is becoming dominated by scrub is said to be 'scrubbing-over'.

- **SCS**
Somatic Cell Score is calculated from the Somatic Cell Count (SCC). When milk is produced, a small number of cells are also transferred to the milk (along with the proteins, fat, water, and minerals that make up milk). Although all milk contains some of these cells, milk quality is affected if they are present in very high numbers. Milk processors limit the amount they will allow in milk they buy from farmers. Also, knowing the SCS for an individual cow can help the farmer tell if the cow is healthy because irritation in the udder can cause higher SCS. Health management has the biggest effect on SCS, but just like some people inherit a higher chance of getting ear infections, cows can inherit traits which cause higher SCS. Next to traits like milk or protein production, SCS has a low heritability.

- **scurs**
regrowth of horn tissue after dehorning.

- **secondary tumour**
cancer that has spread from the organ in which it first appeared to another organ. For example, breast cancer cells may spread (metastasise) to the lungs and cause the growth of a new tumour. When this happens, the disease is called metastatic breast cancer and the new tumour is called a secondary tumour.

- **secondary waste treatment**
usually, a second treatment of waste water to reduce pollution potential or health hazard. With animal manure an example would be storage in an anaerobic treatment lagoon and sprinkler field application.

- **secondary woodland**
woodlands that occupy sites that have not been continuously wooded since 1600 AD See **ancient woodland**.

- **second-look surgery**
surgery performed after primary treatment to determine whether tumour cells remain.

- **secretary**
group leader with specific tasks.

- **sectioned and formed**
(or Chunked and Formed) A boneless ham that is made from different cuts, tumbled or massaged and reassembled into a casing or mold and fully cooked. During this process it is usually thoroughly defatted.

- **sectoral**
relating to one or more sectors (e.g. sectors of society).

- **sedentary**
tending to stay in one place.

- **sedge**
a member of the plant family Cyperaceae.

- **sediment loading**
the amount of sediment (grains of rock material) held in suspension in water such as that in a river or other type of waterbody.

- **seed bank**
the ungerminated but viable seeds that lie in the soil. An artificial store in which seeds are kept usually as a means to ensure that particular species may be conserved.

- **seed productivity**
the amount of viable seed that is produced by an individual plant.

- **seed screening**
removal of unwanted seed from material to be sown as a crop.

- **seed wastage**
seeds that are not harvested (e.g. corn that is spilt as the combine harvester passes during harvest).

- **seepage**
a slow leak.

- **selection**
the causing or allowing of certain individuals in a population to produce the next generation. Artificial selection is that practiced by man. natural selection is that practiced by nature.

- **selection intensity**
the margin of true genetic superiority of those animals selected in comparison to all those from which the choices were made.

- **selective medium**
nutrient material constituted such that it will support the growth of specific

organisms while inhibiting the growth of others.

- **self-feeder**
  a feeding system that allows animals to eat at will.
- **sell by date**
  a calendar date on the packaging of a food product that indicates the last day the product can be sold.
- **semen**
  the male reproductive fluid containing sperm.
- **semi-structured interviewing**
  an informal guided dialogue.
- **sensitivity tests**
  tests used to determine the most effective method of treatment of disease by testing the resistance of the microorganism to classes of antibiotics.
- **sentinel lymph node**
  the lymph node closest to the primary tumour. Cancer cells may appear first in the sentinel node before spreading to other lymph nodes.
- **separator**
  formerly a centrifuge device used to remove the fat from milk on the farm, but now used primarily at processing plants.
  A device used to separate manure into solids and effluent and accomplished by trickling manure over a sloped screen or mechanically forcing through a screen.
- **sepsis**
  1. a serious disease condition in which live bacteria are in the bloodstream. Can lead to an overwhelming infection and death if not treated properly and rapidly. Also called urosepsis when caused by a urinary infection.
  2. the presence in the blood or other tissues of pathogenic microorganisms or their toxins, the condition associated with such presence.
- **septicemia**
  the presence of microorganisms and their associated poisons in blood (commonly called blood poisoning). If the microorganisms are bacteria, the condition is bacteremia.

- **sequencing**
  decoding a strand of DNA or gene into the specific order of its nucleotides: adenine, cytosine, guanine and thymine. This analysis can be done manually or with automated equipment. Sequencing a gene can require analysing an average of 40,000 nucleotides.

- **sequential**
  one treatment after the other.

- **serology**
  study of blood serum and reactions between the antibodies and antigens therein.

- **serotype**
  the type of microorganism as determined by the kind and combination of constituent antigens associated with the cell.

- **service**
  a term commonly used in animal breeding, denoting the mating of male to female. Also called serving or covering.

- **set-aside**
  a form of land use chosen by farmers that allows them to claim 'arable area payments'. It provides an opportunity to introduce management practices to benefit a range of wildlife.

- **settled**
  a term commonly used to indicate that an animal has become pregnant.

- **settling basin**
  any area that reduces flow velocity and allows particulates to settle from a liquid suspension. With manure suspensions, usually built to allow solids removal on a regular basis.

- **settling pond**
  a manure pit where the flow rate of liquid manure is slowed to allow suspended materials to collect at the bottomed, where they can later be removed.

- **sezary syndrome**
  a form of cutaneous T-cell lymphoma, a cancerous disease that affects the skin.

- **shale**
  a type of rock that splits easilyalong the plane in which it has been laid down.

- **shared facility**
  a building or area in which several types of products are created.

- **shared line**
  equipment that is used to make several types of products. Between batches of different products the machinery is usually sanitised.

- **sharp cheese**
  cheese that is cured over six months and has a richer flavour than mild or medium cheese.

- **sheep dip**
  a liquid disinfectant and insecticide in which sheep are immersed to protect them from pests and diseases.

- **shelf life**
  1. time during which food can be stored under specified conditions without apparent undesirable changes.
  2. the length of time after processing that the product will retain its quality.
  3. the time after processing during which a product remains suitable for human consumption, especially the time a food remains palatable.

- **shrink wrapping**
  plastic film that shrinks when heated, producing a tight, neat fit.Tthe most popular form of grocery store meat packaging is PVC wrapping with foam trays.

- **shrinkage**
  a term used to indicate the amount of loss in body weight, as in dairy steers, when exposed to various conditions and/or slaughter. Also, the decrease in volume of dairy products during storage and the loss of milk or milk solids in processing.

- **shy breeder**
  a male or female of any domesticated livestock that has a low reproductive efficiency.

- **sialyl tn-klh**
  a vaccine composed of a substance that enhances immunity plus an antigen commonly found on some tumours of the colon, breast, lung, ovary, pancreas and stomach.

- **sib (sibling)**
  in genetics, a brother or sister.

▪ **sickle-hocked**
describes an animal having a crooked hock, which causes the lower part of the leg to be bent forward out of a normal perpendicular straight line.

▪ **silage**
1. partially fermented conserved fodder.
2. grass cut during spring and stored in large covered stacks. Fed to stock when there is not a good supply of fresh grass.
3. a feed prepared by chopping green forage (e.g. grass, legumes, field corn) and placing the material in a structure or container designed to exclude air. The material then undergoes fermentation, retarding spoilage. Silage has water content of between 60 and 80%.

▪ **silage (ensilage)**
green forage, such as grass or clover, or fodder, such as field corn or sorghum, that is chopped into a silo, where it is packed or compressed to exclude air and undergoes an acid fermentation (lactic and acetic acids) that retards spoilage.

▪ **silage bags**
large plastic tubes in which forages are fermented. Plastic is removed and discarded as the ensiled feed is fed.

▪ **silo**
1. a storage facility for silage. Usually referring to upright concrete or fiberglass silos.
2. a vertical cylindrical structure, pit, trench, or other relatively airtight chamber in which chopped green crops, such as corn, grass, legumes, or small grain and other livestock feeds are fermented and stored. See **silage**.

▪ **silviculture**
the management of whole woodlands or forests for timber and/or other wood products.

▪ **single-cell protein**
cells or protein extracts from microorganisms, grown in large quantities for use as protein supplements.

▪ **sinistral**
left handed. (The way in which the shells coil in some snail species, for example, may be termed 'sinistral').

▪ **sire**
1. father or male parent in a pedigree.
2. the male parent. The verb means to father or beget.

▪ **sire proving scheme**
livestock Improvement's programme to progeny test bulls.

▪ **sire selection**
process of identifying bulls to be used as service sires with the goal of increasing the genetic potential of the herd.

▪ **sirolimus**
a drug used to help prevent rejection of organ and bone marrow transplants by the body.

▪ **skear**
a rocky or stony outcrop (elsewhere called a 'sacr') among the mudflats and sandflats of estuaries to which oysters may be anchored.

▪ **sketch (photo)**
a method of permanent identification to be cross-referenced with visible identification. Accurate sketches or photographs of Holsteins, Ayrshires and Guernseys provide unique identification.

▪ **skim milk (non-fat milk)**
1. milk from which sufficient cream has been removed to reduce its milk fat content to less than 0.5 percent (usually less than 0.1 percent). Skim milk contains as much protein, lactose, minerals, and water-soluble vitamins and only half as many calories as whole milk. Skim milk is practically cholesterol-free.
2. contains less than 0.5% milkfat and at least 8.25% nonfat solids. Skim milk, which contains at least 10% nonfat solids, must be labeled 'protein fortified' or 'fortified with protein.' Vitamin A is added to all skim milk.

▪ **skin testing**
testing for an immune response to a compound by

placing it on or under the skin.

- **slip**
  to abort.

- **slotted floor**
  a concrete floor design in which slats are positioned in the floor so that cow traffic may work manure through the slats and into a pit beneath the floor of the barn.

- **slurry**
  liquified animal manure.

- **smoke flavouring**
  after curing, some hams are smoked. Smoke flavouring (or smoked) is a process by which ham is hung in a smokehouse and allowed to absorb smoke from smoldering fires. This gives added flavour and colour to meat and slows the development of rancidity.

- **sn-117m dtpa**
  a radioactive chemical used to treat bone pain associated with cancer.

- **SNF**
  Solids Non Fat.

- **solids-not-fat of milk**
  proteins, lactose, minerals, and other water-soluble constituents. This is the same as NFS and NMS.

- **soakway**
  a place where water soaks away into the ground.

- **sodium**
  one of the main ingredients of table salt. Excessive salt or sodium in the diet can increase urinary calcium excretion and block the effects of some corrective medication.

- **sodium nitrite**
  used alone or in conjunction with sodium nitrate as a colour fixative in cured meat and poultry products (bologna, hot dogs, bacon). Sodium Nitrate helps prevent the growth of Clostridium botulinum, which can cause botulism in humans.

- **soft tissue**
  soft tissue refers to fat or muscle.

- **soilage**
  freshly cut green forage often fed to animals in drylot. Also called green chop.

- **solid tumour**
  cancer of body tissues other than blood, bone marrow or the lymphatic system

- **solitary**
  not living in colonies (e.g. some wasps or bees build their own nests separate from members of their own species).

- **solvent-extracted**
  fat or oil removed from materials (such as soybean seeds) by organic solvents.

- **somatic cell count (SCC)**
  1. a measurement of the number of somatic cells present in a sample of milk. A high concentration of more than 500,000 somatic cells per mm of milk indicates an abnormal condition in the udder.
  2. the number of white blood cells per mm of milk or measurement of the number of somatic cells present in a sample of milk. A high concentration of more than 500,000 somatic cells per mm of milk indicates abnormal condition in the udder. This serves as an indicator of mastitis infection when elevated above 200,000.

- **somatic cell gene therapy**
  somatic cell gene therapy involves the insertion of genes into cells for therapeutic purposes; for example, to induce the treated cells to produce a protein that the body is missing. It does not affect genetic makeup of a patient's offspring and generally does not change all, or even most, cells in the recipient. Somatic cell gene therapy is only one way of applying the science of genomics to improve health care.

- **somatic cell score**
  a logarithmic representation of the SCC, often referred to as linear scores because they are linearly related to milk production loss.

- **somatic cells**
  1. cells other than sex or germ cells.
  2. the combination of the leukocytes (white blood cells) from blood and the epithelial cells from the secretory tissue of the udder which indicate the presence of infection or injury in the animal.

3. the cell content of milk is composed of approximately 95 percent leukocytes (white blood cells) from the blood and 5 percent epithelial cells from the secretory tissue of the udder. Leukocytes are present in response to infection or injury, and epithelial cells are present as a result of infection or injury. Collectively, these cells are called somatic cells.

- **somatostatin**
  a substance that stops the release of certain hormones.

- **somatotropin**
  a protein hormone produced by the pituitary gland, which stimulates growth of muscle, bone and mammary development in young animals and increases milk production in lactating animals by making available nutrients for milk synthesis and secretion.

- **souse**
  seasoned and chopped pork trimmings.

- **sow**
  a female hog that has produced at least one litter.

- **spasm**
  an abrupt and forceful contraction of a muscle, usually maintained for several minutes or hours and frequently associated with marked pain.

- **spawn**
  the jelly-like mass of eggs laid by amphibians or fish or the act of laying such eggs.

- **spayed**
  to have surgically removed the ovaries of a female.

- **species**
  a unit used in the classification of plants and animals. It is a group of organisms that can interbreed to produce fertile offspring. Where such interbreeding has not been observed (e.g. in fossils, plants that increase in numbers by asexual reproduction) species are defined according to observable similarities between individuals.

- **species action plan**
  a 10-15 year plan which sets objectives and targets for

the maintenance or enhancement of their populations and range, and the actions necessary to achieve them.

- **species recovery programme**
  an english nature-led initiative to conserve some of the country's most threatened wildlife by, among other methods, reintroduction of species to sites where they have gone extinct.

- **species richness**
  the number of species in an ecosystem.

- **sperm**
  the mature male sex cell of an animal by which the ovum is fertilised.

- **sphincter**
  a ring-shaped muscle that allows an opening to close tightly, such as the sphincter muscle in the lower end of a cow's teat.

- **splicing**
  the removal of introns and joining of exons to form a continuous coding sequence in RNA.

- **spore**
  a general term for small (often microscopic) reproductive units consisting of one or a few cells. Ferns and bacteria, for example, produce spores.

- **spring**
  a type of FEN comprising the vegetation associated with an upwelling of water from the land surface which often occur at the head of flushes and water courses.

- **springer**
  a term commonly associated with female cattle showing signs of advanced pregnancy. Often used to refer to heifers close to term with their first calf.

- **squamous cell**
  a layer of cells that covers internal and external surfaces of the body.

- **ß -casein**
  (molecular weight 24,000; 209 residues, 35 prolines) Highly charged N-terminal region and a hydrophobic C-terminal region. Very amphiphilic protein acts like a detergent molecule. Self-association is temperature dependant, will form a large polymer at 20°C but not at 4°C. Less sensitive to calcium precipitation.

- **ß -lactoglobulins**
(MW - 18,000; 162 residues) This group, including eight genetic variants, comprises approximately half the total whey proteins. ß -Lactoglobulin has two internal disulphide bonds and one free thiol group. The conformation includes considerable secondary structure and exists naturally as a noncovalent linked dimer. At the isoelectric point (pH 3.5 to 5.2), the dimers are further associated to octamers but at pH below 3.4, they are dissociated to monomers.

- **stable disease**
cancer that is not decreasing or increasing in scope or severity.

- **stag**
  1. a male animal castrated at or near maturity after having developed definite masculine characteristics.
  2. the extent of a cancer, especially whether the disease has spread from the original site to other parts of the body. Staging refers to the determination of the extent of cancer.

- **stage 11 hodgkin's disease**
cancer is found in two or more lymph node areas on the same side of the diaphragm (the thin muscle under the lungs that helps one breathe), or cancer is found in only one area or organ outside of the lymphatic system and in the lymph nodes around it. Other lymph node areas on the same side of the diaphragm may also have cancer.

- **stage 11 non-hodgkin's lymphoma**
cancer is found in two or more lymph node areas on the same side of the diaphragm (the muscle under the lungs that helps breathing). Cancer is found in only one area or organ outside the lymph nodes and in the lymph nodes around it. Other lymph node areas on the same side of the diaphragm may also have cancer.

- **stage i hodgkin's disease**
cancer is found in only one lymph node area or in only one area or organ outside the lymph nodes.

■ **stage i non-hodgkin's lymphoma**
cancer is found in only one lymph node area or in only one area or organ outside the lymph nodes.

■ **stage iii hodgkin's disease**
cancer is found in lymph node areas on both sides of the diaphragm (the thin muscle under the lungs that helps one breathe). The cancer may have also spread to an area or organ near the lymph node areas and/or to the spleen.

■ **stage iii non-hodgkin's lymphoma**
cancer is found in lymph node areas on both sides of the diaphragm. The cancer may also have spread to an area or organ near the lymph node areas and/or to the spleen.

■ **stage iv hodgkin's disease**
cancer has spread in more than one place to an organ or organs outside the lymph system, or cancer has spread to only one organ outside the lymph system, but lymph nodes far away from that organ are involved. Cancer cells may or may not be found in the lymph nodes near these organs.

■ **stage iv non-hodgkin's lymphoma**
cancer has spread to more than one organ or organs outside the lymph system. Cancer cells may or may not be found in the lymph nodes near these organs. Cancer has spread to only one organ outside the lymph system, but lymph nodes far away from that organ are involved.

■ **stage of lactation**
period of milk production during a lactation determined by the length of time since parturition.

■ **staghorn**
the branched shape of certain large stones. Usually associated with struvite or infection stones, the actual stone material can be almost any chemical stone ingredient.

■ **stale**
a period when an animal does not work or lactate at the normal standards, as opposed to bloom. Also refers to milk products that

have deteriorated in storage.

- **stall**
a cow housing cubicle.

- **stanchion**
  1. a device with two rails that was closed around a cows neck after she entered a stall and to keep her restrained in the stall.
  2. a specially designed headgate to hold an animal in place while allowing feeding and resting.

- **stand**
a sizeable area of vegetation comprised of the same species (e.g. a stand of Bracken).

- **standard plate count**
  1. a test used to measure the live bacteria present in a sample of milk.
  2. determination of the degree of bacterial contamination of a sample made by enumeration of the colonies appearing in a plate.

- **standardisation**
  1. making milk with constant butterfat through partial skimming.
  2. the process of adjusting the milkfat and solids-non-fat content of milk to meet a required standard.

- **standby pool**
an arrangement between cooperatives in markets that have periods of milk shortages and cooperatives in areas of surplus milk production. Cooperatives in the deficit market make monthly payments to certain cooperatives in the surplus regions to assure that they have sources of milk available at reasonable prices when local milk does not supply their Class I needs.

- **staphylococcus aureus**
bacteria of genus Staphylococcus, which are pathogenic.

- **starter culture**
live bacterial culture purposely inoculated into a food such as milk and which contributes a useful conversion process, for example, milk is converted by starter cultures into yogurt.

- **statistically significant**
it usually refers in research to tests for differences resulting from treatments.

The reliability of such differences is expressed, as degree of probability or the percentage of time an observation would be expected to fall outside a certain range of variation from normal observation.

- **statutory**

  a legal requirement. For example, a 'statutory consultee' is a body that must be consulted because some provision in law.

- **statutory agency**

  a government body the powers of which are defined in law (e.g. English Nature, Environment Agency, countryside agency).

- **steam pasteurisation**

  a technology that uses heat to control or reduce harmful microorganisms in beef. This system passes freshly slaughtered beef carcasses that are already inspected, washed, and trimmed, through a chamber that exposes the beef to pressurised steam for approximately 6 to 8 seconds. The steam raises the surface temperature of the carcasses to 190° to 200° F (88° to 93° C). The carcasses are then cooled with a cold-water spray. This process has proven to be successful in reducing pathogenic bacteria, such as E. coli O157:H7, Salmonella, and Listeria, without the use of any chemicals.

- **steer**

  1. a bovine male castrated before reaching sexual maturity.
  2. a vasectomised or castrated bull.
  3. a male bovine castrated before development of secondary sex characteristics.

- **steinstrasse of**

  literally 'street of stones'. This refers to the collection stone fragments that have lined up in the ureter and have not passed. May require additional surgery to remove. Usually happens after ESWL with large stones when a stent is not used.

- **stent**

  a thin, slender tube designed to bypass any obstruction of the ureter such as from a stone or scar tissue. One end of the stent is placed in the kidney and the other is in the bladder. Each end of the stent usually forms a small circle to help keep it in place. Used to help drain the kidney, by-pass a stone and relieve pain. Also used as a land-mark to help locate smaller or hard to see stones. Helps protect against clogging when large stones, over 1 cm in size are fragmented.

- **step-up parlour**
  cows step onto raised plat-forms for milking. The milk-ing units are attached from the side.

- **steric stabilisation**
  as already noted, the hairy layer interferes with inter-particle approach.

- **sterile**
  clean, free of any living or-ganisms. Also means unable to reproduce.

- **sterilise**
  to remove or kill all living organisms. Also, to make barren or unproductive, as a vasectomy in bulls.

- **steroid therapy**
  treatment with corticoster-oid drugs to reduce swelling, pain, and other symptoms of inflammation.

- **stillborn**
  born lifeless, dead at birth.

- **stock recovery**
  a firm's removal or correc-tion of product that has not been marketed or that has not left the direct control of the firm. For example, prod-uct is located on premises owned by, or under the con-trol of, the firm, and no por-tion of the lot has been re-leased for sale or use.

- **stocker calves**
  young, growing animals on pasture with very little other feed, with the inten-tion of increasing weight and maturity before being placed in a feedlot.

- **stocking rate**
  the number of livestock per unit area of land.

- **stocks**
  means inventory of sugar on hand at the beginning and at the end of the calen-dar month for which data are being reported.

- **stoke's equation**
  the velocity at which a sphere will rise or fall in a

liquid varies as the square of its diameter. For example, a fat globule with a diameter of 2 microns will rise 4 times faster than a fat globule with a diameter of 1 micron.

- **stone basket**
one of several types of wire instruments used to capture and remove stones in the ureter. Normally used either with fluoroscopy or with special telescopes called ureteroscopes.

- **stop codon**
one of three codons in messenger RNA that signal the end of the amino acid chain in protein synthesis.

- **stover fodder**
mature cured stalks of grain from which seeds have been removed, such as stalks of corn without ears.

- **straws of semen**
fine plastic straws used to store semen.

- **streak canal**
small canal located in the end of each teat, through which the milk passes immediately prior to expulsion. Also called the teat meatus.

- **streptavidin**
a bacterial protein that is used in a two-step process of treating patients with radiolabeled biotin (a vitamin). A monoclonal antibody that finds tumour cells is given with streptavidin that attaches to the radioactive biotin.

- **stricture**
the medical term for a narrowed or scarred area. Often occurs in the urinary system after instrumentation or stone passage. If there is a stricture of the ureter, any stones trying to pass will probably get stuck there. Stents work well to gently stretch open strictures of the ureters, but cannot help strictures of other bodily organs.

- **string**
a group of animals within a larger group, or herd.

- **strip cup**
a small cup or device to collect forestrippings and which makes abnormal milk easier to observe.

- **strontium**
a metal often used in a radioactive form for imaging

tests or as a treatment for cancer.

- **strontium chloride sr 89**
  an injectable, radioactive compound that is absorbed by the bone and used to relieve cancer pain.

- **strontium-89**
  an injectable, radioactive compound that is absorbed by the bone, and used to relieve cancer pain.

- **structural funds**
  measures to aid economic and social development in the EU. Objective 5b areas currently cover large tracts of the English uplands but will be replaced by smaller Objective 1 and 2 areas under AGENDA 2000.

- **structural gene**
  a gene that codes for a protein, such as an enzyme.

- **struvite**
  another name for stones caused by infection. Also called 'Triple Phosphate' stones.

- **stubble**
  the short stalks left behind in a field once the crop has been harvested.

- **stud**
  a unit of selected animals kept for breeding purposes (e.g., bulls).

- **su101**
  an anticancer drug.

- **subclinical**
  1. a disease condition without symptoms but often resulting in decreased production or impaired milk quality.
  2. a disease condition without clinical manifestations.

- **subcutaneous**
  beneath the skin.

- **subcutaneous port**
  the placement of a catheter into a vein to receive fluids, including chemotherapy drugs, that will be administered repeatedly or over a period of time.

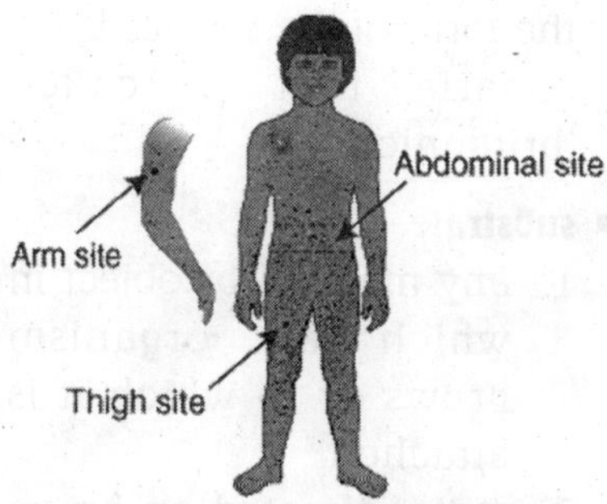

Figure 7: Common sites for subcutaneous injections in children.

- **submerged vegetation**
  plants rooted to the bed of a water body and either completely submerged or with only part of their shoots floating or emergent.

- **sub-species**
  a unit in the classification of plants and animals that subdivides species. Individuals of the same sub-species have certain characteristics in common with each other and tend to interbreed more readily than with other members of the species to which they belong. 'forms' and 'varieties' are terms that have been used that mean the same thing. The scientific name for a given sub-species comprises three parts. Foe example 'Lycia zonaria britannica' denotes the form of the species Lycia zonaria that is called 'britannica'.

- **substrate**
  1. any material or object in which an organism grows or to which it is attached.
  2. material acted on by an enzyme.

- **subway**
  an area beneath the milking pit that houses milk meters, pipelines, vacuum lines and transfer tanks to reduce noise and improve the milker's ability to move around in the pit.

- **succession**
  a gradual sequence of changes in vegetation over a period of time until an equilibrium has been attained and a climax community is established, e.g. ungrazed grassland developing into scrub and then woodland. See **plant community**.

- **succulence**
  a condition of plants characterised by juiciness, freshness and tenderness, making them appetising to animals.

- **sucrose**
  1. means a disaccharide having the chemical formula $C_{12}H_{22}O_{11}$.
  2. sugar derived from sugar cane, sugar beets and sorghum.

- **sugar**
  means any grade or type of saccharine product which is

processed, directly or indirectly, from sugarcane or sugar beets (including sugar produced from sugar beet or sugarcane molasses) and consisting of, or containing, sucrose or invert sugar, including all raw sugar, refined crystalline sugar, liquid sugar, edible molasses, sugar syrup and cane syrup.

- **sugar beet processor**
  means a person who commercially produces sugar directly or indirectly from sugar beets (including sugar produced from sugar beet molasses). The same person may be both a 'sugar beet processor' and either a 'cane sugar refiner' or 'sugarcane processor', or both.

- **sugar cured**
  a term that may appear on ham labels if cane or beet sugar is at least half the sweetening ingredients used and if the sugar is used in an amount sufficient to flavor and/or affect the appearance of the finished product. Most hams contain sugar in the curing mixture.

- **sugar syrup**
  means a direct-consumption sugar with a sucrose or sucrose- equivalent invert sugar content of less than 94 percent of the total soluble solids.

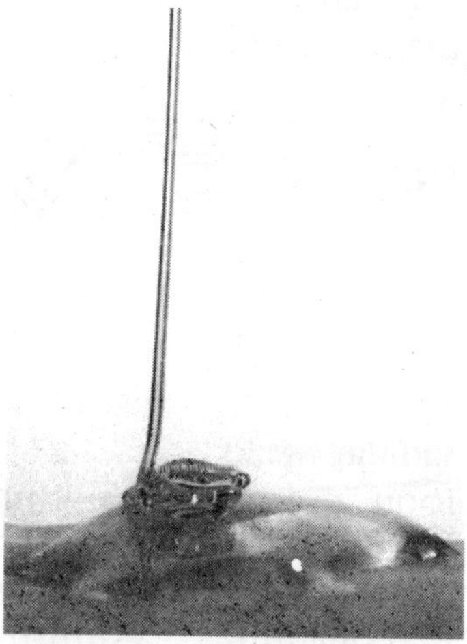

- **sugarcane processor**
  means a person who commercially produces sugar directly or indirectly from sugarcane (including sugar produced from sugarcane molasses). The same person may be both a 'sugarcane processor' and a 'cane sugar refiner' or 'sugar beet processor', or both.

- **suicide gene**
  a gene that codes for an antibiotic that can kill the

host bacterial cell. It is genetically modified into the bacterium along with a molecular switch that is controlled by a nutrient in the environment. When the nutrient disappears, the suicide gene is switched on and the bacterium dies.

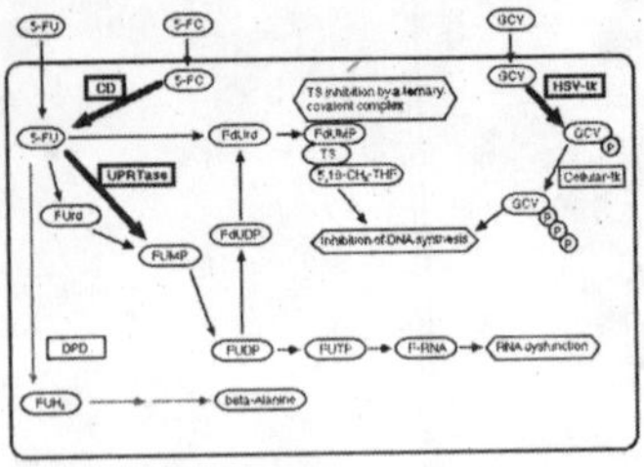

- **sulfiting agents**
food preservatives composed of potassium or sodium bisulfite or potassium metabisulfite. Sulfiting agents are used in processing of beer, wine, baked goods, soup mixes, and some imported seafoods and by restaurants to impart a 'fresh' appearance to salad fruits and vegetables. The chemicals can cause a severe allergic reaction in people who are hypersensitive to sulfites. The reactions are marked by flushing, faintness, hives, headache, GI distress, breathing difficulty, and in extreme cases, loss of consciousness and death.

- **sulindac**
a nonsteroidal anti-inflammatory drug (NSAID) being tested in clinical trials for its ability to prevent cancer.

- **superhutches**
calf housing structures, often open on one side, designed for a small number of calves when first grouped immediately after weaning.

- **supplement**
to add minerals, vitamins, or other minor ingredients (volumewise) to a ration.

- **supplementary feed**
feed used to supplement livestock dietary requirements usually during the winter. This often consists of conserved fodder, such as hay or silage, feed blocks or concentrates.

- **supplementary feed**
food (silage and/or hay) fed to 'supplement' a grass-based diet.

- **supplementation**
adding nutrients to the diet, usually in high doses.

- **suppressor gene**
  a gene that can reverse the effect of a mutation in other genes.
- **supraclavicular lymph nodes**
  lymph nodes located above the clavicle or 'collar bone'.
- **supratentorial**
  located in the upper part of the brain.
- **surface water**
  water standing or flowing on the surface of land (Compare with groundwater).
- **surplus milk**
  the quantity of grade A fluid milk in excess of that needed for Class I purposes.
- **surprise tests**
  these are tests for verification performed on cows whose projected 2X (twice daily milking), 305D (305-day lactation), ME (mature equivalency) records exceed certain levels of milk and/or fat as established by breed associations.
- **surveillance**
  a system of monitoring the health of the population, which is used to prevent foodborne illness outbreaks from increasing.
- **survey**
  a tool used by epidemiologists to understand the state of health of the population or to identify the source of a foodborne outbreak.
- **sustainability indicator**
  a measure of sustainable development.
- **sustainable development**
  development that meets the needs of the present without compromising the ability of future generations to meet their own needs.
- **swamp**
  a broad term covering mostly tall, emergent vegetation occurring adjacent to open water.
- **sward**
  above ground components of grassland vegetation comprising grasses and herbs.
- **swarming behaviour**
  behaviour observed in some social insects when they group together into 'swarms' often associated with mating and/or moving to found new colonies.
- **sweet butter**
  unsalted butter.

- **sweetened condensed milk**
a canned whole milk concentrate containing at least 28.0% milk solids and 8.5% milkfat It is pasteurised and may be homogenised. The nutritive sweetened added to it helps prevent spoilage. Sweetened condensed milk cannot replace evaporated milk in recipes because of its high sugar content.

- **swing parlour**
parlour characterised by having the milking units positioned in the middle of the parlour for use by cows on both sides.

- **Swiss-style yogurt**
plain or flavoured yogurt with fruit, blended throughout.

- **switch**
the brush of hair on the end of a bovine's tail.

- **SWOT analysis**
Strengths, Weaknesses, Opportunities and Threats analysis.

- **syrup**
means a viscous, concentrated sugar solution resulting from the evaporation of water, or the remaining liquor after crystallisation of sugar from a solution.

- **systemic**
affecting the entire body.

- **systemic therapy**
treatment that reaches and affects cells throughout the body.

- **systems biology**
a hypothesis-driven field of research that creates predictive mathematical models of complex biological processes or organ systems.

- **t cells**
one type of white blood cell that attacks virus-infected cells, foreign cells, and cancer cells. They also produce a number of substances that regulate the immune response.

- **t lymphocytes (t-cells)**
white blood cells that are produced in the bone marrow but mature in the thymus. They are important in the body's defense against certain bacteria and fungi, help B lymphocytes make antibodies and help in the recognition and rejection of foreign tissues. T lymphocytes may also be important

in the body's defense against cancers.

- **table cream**
  see **light cream**

- **tacrolimus**
  a drug to suppress the immune system.

- **tail bands**
  used for marking treated cows. Rubber bands are sometimes used to dock cow's tails.

- **take**
  to accept a male in coitus. Also a lay term meaning to become pregnant.

- **take-off**
  see **automatic take-off**.

- **tallow**
  the white nearly tasteless solid rendered fat of cattle and sheep used chiefly in soap, candles and lubricants.

- **tanbark trail**
  a term often associated with those who exhibit animals in competition at fairs and livestock shows.

- **tandem parlour**
  parlour design where cows line up head to tail in individually opening stalls.

- **tankerload**
  in milk collection and distribution, large quantities of milk in a large stainless steel tank borne on a transport truck—the method of transporting milk.

- **tattoo**
  a method of permanent identification to be crossreferenced with visible identification. Permanent ink used with tattoo numbers creates a lifetime record of identification in the ears of dairy cattle.

- **taxon (plural = taxa)**
  a unit of classification used in taxonomy (e.g. 'species').

- **taxonomy**
  the study of the classification of plants and animals according to the differences and similarities between them.

- **t-cell depletion**
  treatment to destroy T cells, which play an important role in the immune response. Elimination of T cells from a bone marrow graft from another person may reduce the chance of an immune reaction against the patient's own tissues.

- **t-cell lymphoma**
  a disease in which certain cells of the lymph system (called T lymphocytes) become cancerous (malignant).

- **teaser**
  castrated or vasectomised steer used to tease, or sexually excite, artificial breeding bulls so semen can be collected.

- **teat**
  1. the appendage on the udder through which milk from the udder flows.
  2. a small protuberance or appendage on the udder, through which milk from the udder flows.

- **teat dip- pre and post-milking**
  substance that kills bacteria and helps to seal the teat end to prevent entry of bacteria into the udder between milkings. May contain emollients to improve teat end condition for use in cold, winter conditions.

- **teat meatus**
  small canal located in the end of each teat, also called a streak canal.

- **technetium tc 99m sulphur colloid**
  a radiolabeled substance that is used to help identify sites of tumour development.

- **technology transfer**
  the process of transferring discoveries made by basic research institutions, such as universities and government laboratories, to the commercial sector for development into useful products and services.

- **tegafur**
  an investigational anticancer drug.

- **temozolomide**
  a chemotherapy drug that interferes with the DNA in cancer cells and prevents them from growing.

- **temperate**
  possessing a moderate climate. Belonging to one of the three broad climate zones that are to be found on earth. Apart from the temperate zone (found generally in middle latitudes) there are the 'torrid zone' (i.e. the Tropics) and the 'frigid zone' (i.e. the Arctic and Antarctic).

- **temperature**
  at 4° C, beta-casein begins to dissociate from the micelle, at 0° C, there is no micellar aggregation, freezing produces a precipitate called cryo-casein.
- **template**
  a molecule that serves as the pattern for synthesising another molecule.
- **temporary pools**
  ponds that dry up at some time during the year (usually in the summer).
- **teniposide**
  a chemotherapy drug that inhibits cancer cell growth by interfering with the formation of DNA.
- **term**
  the gestation period.
- **terminator**
  sequence of DNA bases that tells the RNA polymerase to stop synthesising RNA.
- **tertiary structure**
  the total three-dimensional shape of a protein that is essential to protein function.
- **tertiary waste treatment**
  usually, the final treatment of waste water to reduce pollution potential or health hazard allowing discharge into bodies of water. Not economically practical with animal manure waste waters.
- **test**
  to evaluate the producing ability of a lactating animal, to determine the percentages of various components of milk, such as milk fat, to examine the blood of an animal for specific pathogens.
- **testing of milk**
  see **milk testing**.
- **tetany**
  a condition in an animal in which there are localised, spasmodic muscular contractions.
- **tether**
  to tie an animal with a rope or chain to allow grazing but prevent straying.
- **tethering**
  to tie up or attach to.
- **tetrad**
  a grouping of four adjacent one km squares often used in biological recording. See **record**.

- **thalidomide**

  a drug which, if taken in early pregnancy, may cause birth defects. It is also being studied in clinical trials for the treatment of a number of conditions and diseases, including cancer.

- **thaw**

  to warm, to change from frozen to liquid state.

- **the casein micelle**

  most, but not all, of the casein proteins exist in a colloidal particle known as the casein micelle. Its biological function is to carry large amounts of highly insoluble CaP to mammalian young in liquid form and to form a clot in the stomach for more efficient nutrition. Besides casein protein, calcium and phosphate, the micelle also contains citrate, minor ions, lipase and plasmin enzymes and entrapped milk serum. These micelles are rather porous structures, occupying about 4 ml/g and 6-12% of the total volume fraction of milk.

  The 'casein sub-micelle' model has been prominent for the last several years, and is illustrated and described with the following link, but there is not universal acceptance of this model, and mounting research evidence to suggest that there is not a defined sub-micellar structure to the micelle at all. Another model of a more open structure is also defined with the following link.

  In the submicelle model, it is thought that there are small aggregates of whole casein, containing 10 to 100 casein molecules, called submicelles. It is thought that there are two different kinds of submicelle, with and without kappa-casein. These submicelles contain a hydrophobic core and are covered by a hydrophilic coat which is at least partly comprised of the polar moieties of kappa-casein. The hydrophilic CMP of the kappa-casein exists as a flexible hair.

  The open model also suggests there are more dense and less dense regions within the midelle, but there is less of a well-defined structure. In this model, calcium phosphate nanoclusters bind caseins and provide for the differences in density within the casein micelle.

- **the dipstick**
urolithiasis The disease process in which a stone is formed anywhere in the urinary system. While this usually refers to kidney stones, the term includes bladder stones as well.

- **therapeutics**
compounds that are used to treat specific diseases or medical conditions.

- **thioguanine**
a drug used in the treatment of cancer.

- **thiotepa**
belongs to the group of anticancer drugs called alkylating agents.

- **thorax**
a region of the body. In vertebrates this is the part of the body that contains the heart and lungs. In most arthropods this is the region of the body that lies between the head and the abdomen.

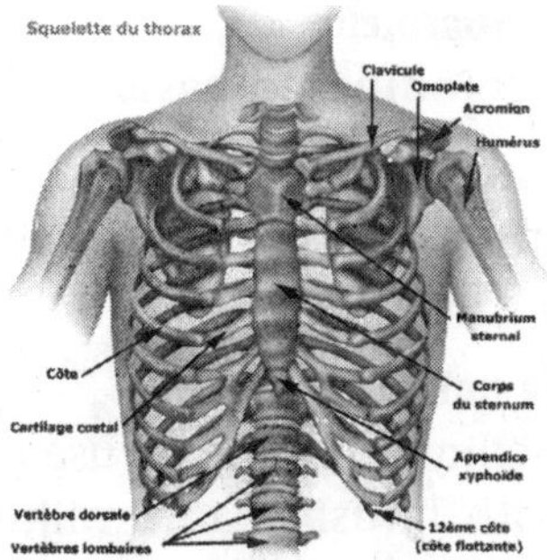

- **thrombocytopenia**
a decrease in the number of platelets in the blood.

- **throughput**
the number of cows that can be milked in a parlour in a given period of time.

- **thymidine**
a chemical compound found in DNA. Also used as treatment for mucositis.

- **thymus**
a lymphoid organ in the lower neck, the proper functioning of which in early life is necessary for development of the immune system.

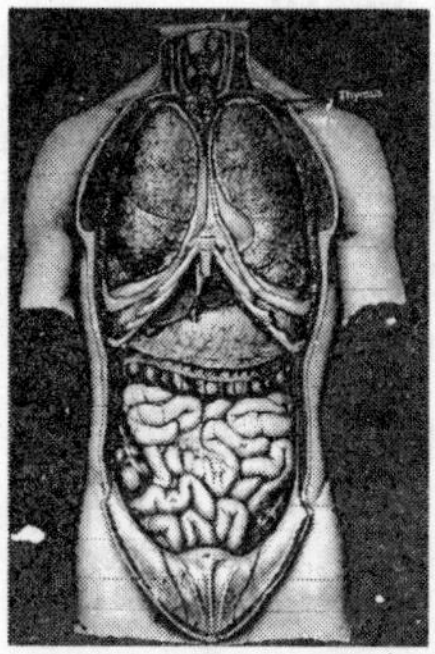

- **thyroprotein**
  see **Iodinated casein.**

- **tiazofurin**
  an anticancer drug used to stop cell growth.

- **tie stall parlour**
  facility is frequently used for both housing and milking. Cows are tied and milked with the cow and operator on the same level.

- **time / area observation**
  counts of numbers/ behavioural observations made over a fixed time in a fixed area.

- **time recording device/monitor**
  a mechanical/electronic device that automatically monitors the interval between milkings. The time is expressed in hours plus minutes, and the device must display the starting and ending time of the previous milking.

- **times milked**
  cows are normally milked twice per day with records being labelled 2x, however, cows may be milked more frequently (3x, 4x, etc.).

- **tin ethyl etiopurpurin**
  a drug used in photodynamic therapy; it destroys cancer cells when activated by light. Also called SnET2

- **tirapazamine**
  a drug that makes tumour cells more sensitive to radiation therapy.

- **tissue culture**
  in vitro growth in nutrient medium of cells isolated from tissue.

- **tissue plasminogen activator (TPA)**
  a protein produced in small amounts in the body that aids in dissolving blood clots.

- **titer**
  the quantity of a substance required to produce a reaction with a given volume of another substance, or the amount of one substance required to correspond with a given amount of another substance. Agglutination titer is the highest dilution of a serum that causes clumping of bacteria.

- **titratable acidity**
  a measure of titaratable hydrogen ions. Includes H+ ions free in solution and those associated with acids and proteins.

- **TMR**
  total mixed ration.

- **tnp-470**
  belongs to a group of drugs known as angiogenesis inhibitors, which block the growth of new blood vessels.

- **toe out**
  to walk with the feet pointed outward. Also called splay-footed.

- **tolerance, pesticide residue**
  the amount of pesticide residue allowed by regulation to remain in or on a food sold in interstate commerce. Whenever a pesticide is registered for use on a food or a feed crop, a tolerance (or exemption from the tolerance requirement) must be established.

- **tom**
  a male turkey.

- **tomograms**
  a type of X-ray where both the X-ray projector and film are revolving around the patient. This technique is able to focus better on the kidney and permit better visualisation even in difficult cases. Plain tomograms without contrast will help demonstrate even relatively small stones not visible with other methods.

- **top**
  to cut grass sward or selected plant species to favour growth of palatable shoots or for weed control.

- **topical**
  at the surface of the body. Topical chemotherapy, for example, is applied to the skin.

- **topography**
  the surface features of the Earth.

- **topotecan**
  belongs to a group of anticancer drugs known as camptothecin analogues.

- **torpor**
  drowsiness often induced by low temperature.

- **total digestible nutrients (TDN)**
  a standard evaluation of the nutritional merit of a particular feed for farm animals which includes all the digestible organic nutrients—protein, fibre, nitrogen-free extract and lipids.

- **total milk solids (TMS)**
primarily milk fat, proteins, lactose and minerals.

- **total-body irradiation**
radiation therapy to the entire body. Usually followed by bone marrow or peripheral stem cell transplantation.

- **toxic**
harmful.

- **toxin**
  1. a poisonous substance that may be found in food.
  2. a poisonous substance produced by certain microorganisms or plants.

- **toxins**
the poisons produced by certain microorganisms. They are products of cell metabolism. The symptoms of bacterial diseases, such as diphtheria, tetanus, botulism, and staphylococcal food poisoning, are caused by toxins.

- **toxoid**
a denatured toxin. It retains the ability to stimulate the formation of antitoxin in an animal's body.

- **tracer**
a substance, such as a radioisotope, used in imaging procedures to diagnose disease or to see how the body is responding to treatment.

- **trade barriers**
rules and regulations that hamper the trade of commodities. Tariffs, fees for imported goods and import limits or quotas are barriers to trade.

- **training**
education of milk producer group members.

- **trait**
a specific feature.

- **transcription**
synthesis of messenger (or any other) RNA on a DNA template.

- **transdifferentiation**
the process whereby a specialised cell de-differentiates and re-differentiates into a different cell type, or the process whereby an adult stem cell from a specific tissue type becomes a cell type from a very different tissue (for example a nerve stem cell differentiates into a kidney cell).

- **transduction**
transfer of genetic material from one cell to another by

means of a virus or phage vector.

- **transfection**
  infection of a cell with nucleic acid from a virus, resulting in replication of the complete virus.
- **transfer RNA (tRNA)**
  RNA molecules that carry amino acids to sites on ribosomes where proteins are synthesised.

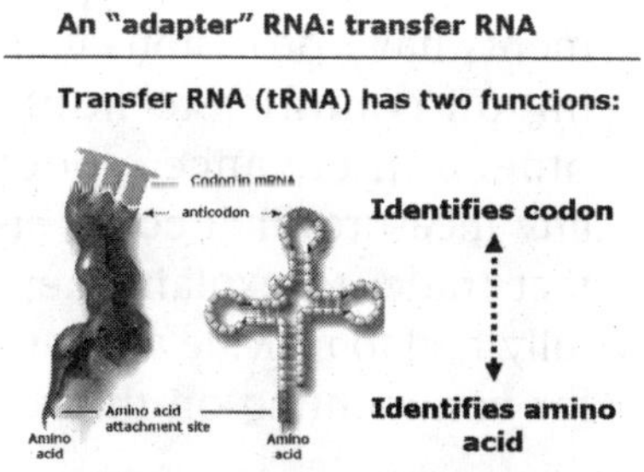

- **transformation**
  change in the genetic structure of an organism by the incorporation of foreign DNA.
- **transfusion**
  the inftision of components of blood or whole blood into the bloodstream.
- **transgenic organism**
  1. an organism formed by the insertion of foreign genetic material into the germ line cells of organisms. Recombinant DNA techniques are commonly used to produce transgenic organisms.
  2. an organism formed by the insertion of foreign genetic material into a germ cell.

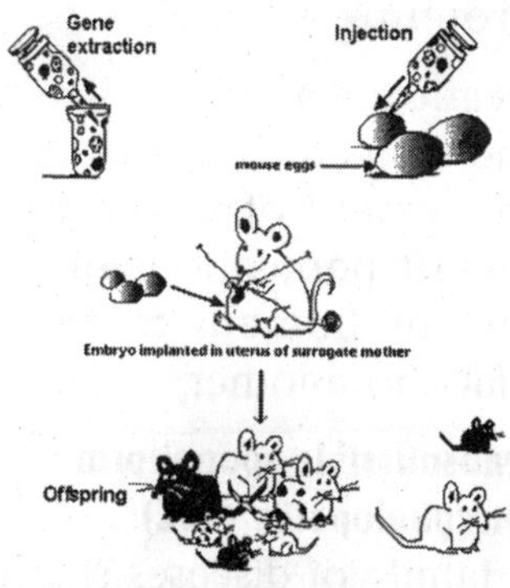

Transgenic Mice

- **transient**
  lasting for a short time only.
- **transition**
  the process of changing from one state to another. In terms of plant communities the term is applied to the zone where one type of vegetation grades into another (e.g. scrub grading into grassland).
- **transition cow**
  a cow that is from 2 months before 1 month after expected calving date.

- **transition housing**
  barns designed especially for transition cows, often including a maternity area.

- **translation**
  process by which the information on a messenger RNA molecule is used to direct the synthesis of a protein.

- **translocation**
  the deliberate and planned movement of whole habitats or populations of species by humans from one place to another.

- **transmissible spongiform encephalopathy (TSE)**
  a family of diseases sharing some common characteristics, including a prolonged incubation period ranging from a few months to years and progressively debilitating neurological illnesses, which are always fatal. Examples of other TSEs include scrapie (sheep and goats), chronic wasting disease (deer and elk), feline spongiform encephalopathy (cats), kuru (humans), Creutzfeldt-Jakob Disease (humans), and variant Creutzfeldt-Jakob Disease (humans).

- **transparency**
  a World Trade Organisation principle stipulating that a country's policies and regulations affecting foreign trade should be clearly communicated to its trading partners. For example, out of recognition that sanitary and phytosanitary measures may (sometimes deliberately) be unclear, arbitrary, or capricious, recent international trading agreements have provisions calling on countries to notify others, in advance, about any measures that could affect trade, to explain them fully, and to provide a means for commenting on them.

- **transposon**
  a segment of DNA that can move around and be inserted at several sites in bacterial DNA or in a phage, thus alerting the host's DNA.

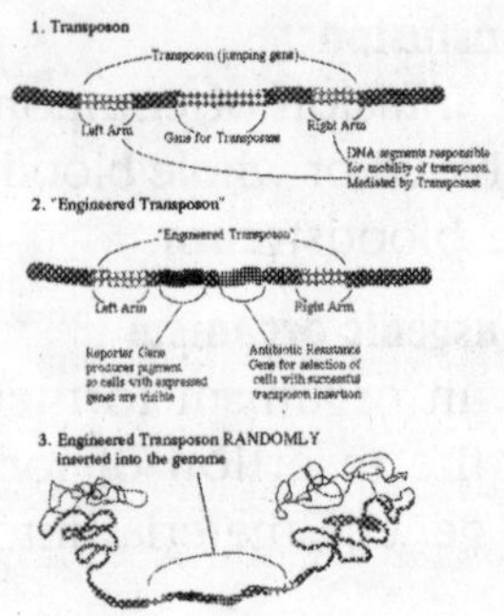

- **treasurer**
  group leader with specific tasks.

- **tretinoin**
  a drug that inhibits the growth of some types of cancer cells.

- **trial**
  a test.

- **tributyrin**
  a drug used to cause cancer cells to mature into normal blood cells.

- **trimetrexate**
  an anti-cancer drug used in the treatment of acute lymphoblastic leukemia that is resistant to methotrexate.

- **trimetrexate glucuronate**
  an anti-cancer drug used in the treatment of acute lymphoblastic leukemia that is resistant to methotrexate.

- **triple phosphate**
  stands for magnesium, ammonium phosphate and calcium phosphate. Another name for struvite stones. The 'triple' stands for the three ingredients of calcium, magnesium and ammonium that are found in these stones.

- **triptorelin**
  belongs to the group of drugs known as luteinising hormone-releasing hormone agonists. Used to block hormone production in ovarian ablation.

- **true protein**
  true protein reflects only the nitrogen associated with protein and does not include the nitrogen from non-protein sources.

- **ts**
  total Solids.

- **tumour infiltrating lymphocytes**
  white blood cells that have left the bloodstream and migrated into a tumour.

- **tumour necrosis factors (TNF)**
  rare proteins of the immune system that appear to destroy some types of tumour cells without affecting healthy cells.

- **tumour-derived**
  taken from a patient's own tumour-tissue; may be used in the development of a vaccine that enhances the body's ability to build an immune response to the tumour.

- **type**
  the physical conformation of an animal.

- **type classification**
  a program sponsored by breed associations whereby a registered animal's conformation may be compared with the 'ideal' or 'true' type animal of that breed by an official inspector (classifier).

- **ucn-01**
  belongs to the group of anticancer drugs known as staurosporine analogues.

- **udder**
  1. glandular organ of cows which contains milk producing cells.
  2. the encased group of mammary glands provided with teats or nipples as in a cow, ewe, mare or sow. Also referred to as a bag.

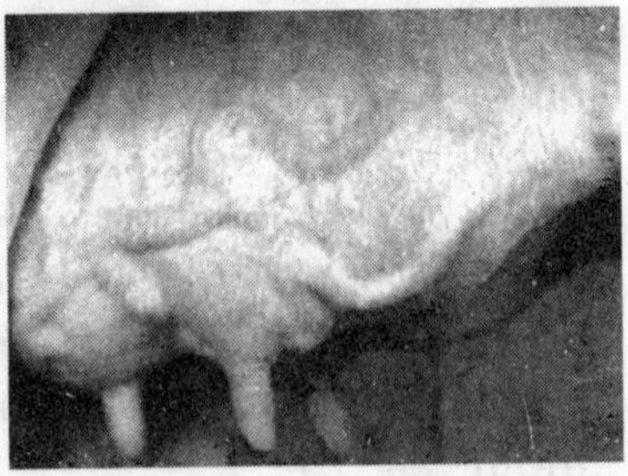

- **uf (ultrafiltration)**
  one of several types of membrane filtration technology. When applied to skim milk, the albumin proteins may pass into the permeate but the casein proteins are captured in the retentate. Can be applied as a pretreatment before cheese making.

- **UHT**
  1. Ultra High Temperature heat treatment for preserving liquid foods such as milk by exposure to brief, intense heating between 135-140 degrees C to destroy unwanted or disease-causing bacteria.
  2. ultra high temperature pasteurisation process where milk is heated to 280° F for at least 2 seconds. Typically, UHT milk is packaged in aseptic containers so as not to require refrigeration.

- **UHT (ultra-high-temperature) pasteurised**
  the process of heating raw milk at extremely high temperatures for a few seconds and then aseptically packaging it so that the milk can be

stored under non-refrigerated conditions and not spoil, After opening, UHT milk must be treated like fresh milk,

- **ultrafiltration**
  extremely fine filtration.

- **ultrapasteurised**
  heat treatment method to preserve the shelf life of milk.

- **ultrapasteurised**
  UHT milk or milk pasteurised nearly at that level but not aseptically packaged. Often used in single serving coffee creamers and some beverage milk for food service.

- **ultra-pasteurised**
  a similar process to UHT where raw milk is heated to extremely high temperatures for a few seconds. Ultra-pasteurised products will keep longer than conventionally pasteurised products, but unlike UHT milk, they must be refrigerated.

- **ultrasonic**
  sound waves with a frequency above the upper limit that is audible to humans.

- **ultrasound**
  high frequency sound waves. Usually used like sonar for diagnosis. Has the advantage of safety. When used as a therapeutic probe, ultrasound can drill through most stones and pulverise them.

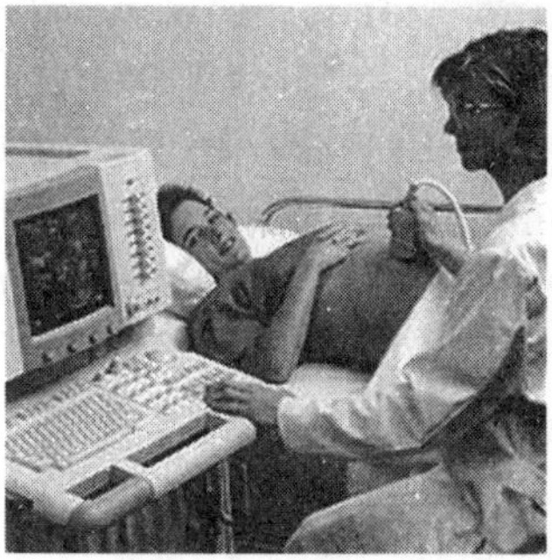

- **ultraviolet radiation**
  a form of high-energy radiation used in the treatment of cancer.

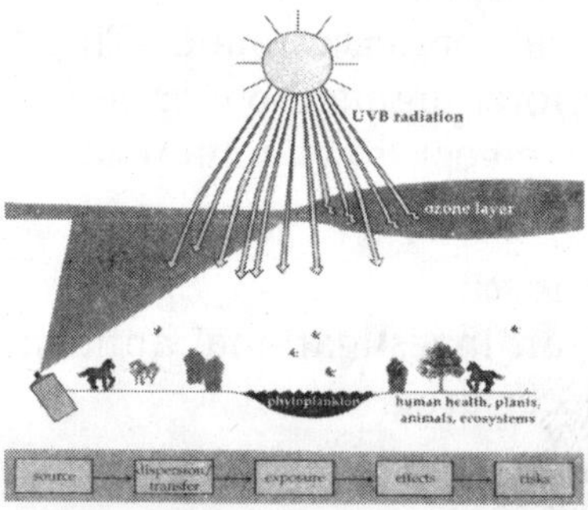

- **umbilical cord blood**
  blood from the placenta (afterbirth) that contains cells that can grow healthy new bone marrow.

- **uncommitted inventories**
  stocks owned by the CCC at a given time which have not been sold, donated or in any way committed for use.

- **unenclosed land**
  land that is not fenced or walled such as much of Moorland/Fell in the county.

- **unimproved grassland**
  grassland that has not been agriculturally improved.

- **unresectable**
  unable to be surgically removed.

- **unsalted butter**
  product often used in cooking and baking. Unsalted butter has a shorter shelf life than salted butter because salt preserves butter. Therefore, unsalted butter should be kept frown until ready to use.

- **uracil**
  an investigational anticancer drug.

- **ureter**
  the name of the muscular tube that empties the kidney and carries urine down to the urinary bladder. Normally it gently squeezes the urine downwards much like the intestines slowly pass food along from the stomach to the rectum.

- **ureteropelvic junction (UPJ)**
  the anatomical location where the inside of the kidney connects to the ureter. It's a common place for strictures and blockages. One of the three most likely locations for a stone to get stuck.

- **ureteroscope**
  a specially made telescope designed to be passed through the urinary bladder into the ureter. The longer, flexible versions can reach all the way up to the kidney.

- **ureteroscopy**
  use of the ureteroscope to examine the ureter.

- **ureterovesical junction (UVJ)**
  the anatomical location where the ureter joins with the urinary bladder. The entry is at an angle so that urine will not return up to the kidney during voiding when the bladder pressure is high. This is the narrowest part of the ureter and where many stones will get stuck.

- **urethra**
  the tube that allows passage of urine from the bladder to the outside.

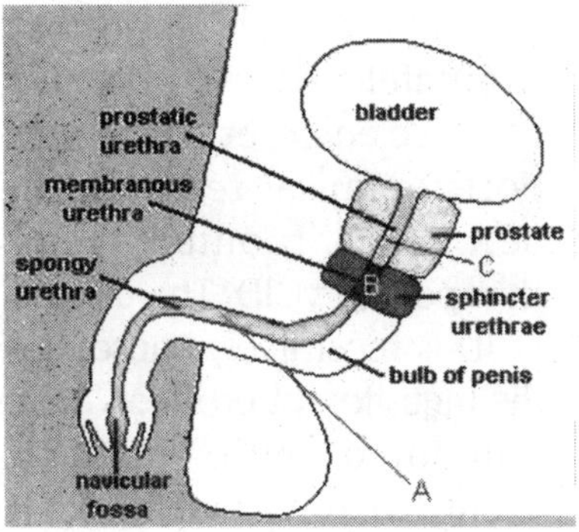

- **uric acid**
  the final chemical endpoint in humans of purine (protein) metabolism. Most of it is excreted in the urine. Uric acid can form stones, make the urine more acidic and increase calcium stone formation.

- **urinalysis**
  an examination of the urine. Usually best done with a chemical dipstick and a microscope, but often done just with.

- **urine**
  the liquid excreted by the kidneys. Normally it has a clear amber color. Urine does not normally contain sugar, albumin, pus, blood, bacteria, acetone, casts or crystals.

- **urogenital tract**
  the urinary and genital organs (kidney, ureter, bladder, prostate, penis, urethra, etc.).

- **urokinase**
  a drug that dissolves blood clots or prevents them from forming.

- **urologist**
  a specialist in the field of urology.

- **urology**
  the medical specialty field that primarily deals with surgical problems affecting the kidneys, urinary bladder and male genitalia. Urology is a surgical specialty.

- **urosepsis**
  a serious infection where the source of the bacteria was originally from the urinary tract.

- **UTI**
  Urinary Tract Infection. Usually refers to a bladder infection.

- **vaccination**
  the process of artificially stimulating the immune response in an animal to a an

altered biological material resulting in re sistance to an infectious disease.

- **vaccine**

1. a preparation that contains an antigen, consisting of whole disease causing organisms (killed or weakened) or parts of such organisms, that is used to confer immunity against the disease that the organisms cause. Vaccine preparations can be natural, synthetic or derived by recombinant DNA technology.
2. a compound or group of compounds designed to produce an immune response to a tumour or disease.

- **vacuum packaging**

rigid or flexible containers from which substantially all air has been removed before sealing. Carbon dioxide or nitrogen may be introduced into the container. This process prolongs shelf life, preserves the flavours and retards bacterial growth.

- **van der waals forces**

no sucess in relating these forces to micellular stability.

- **vCJD variant creutzfeldt-jakob disease**

vCJD is a variant of the most commonly identified TSE in humans, classic (sporadic) Creutzfeldt-Jakob Disease (CJD). Scientific evidence supports a causal relationship between BSE outbreaks in Europe and vCJD. The disease vCJD is most likely caused by the ingestion of products contaminated with the BSE agent. There has never been a case of vCJD that did not have a history of exposure within a country where BSE was occurring. Patients with vCJD have primarily been younger and exhibit clinical signs of the disease longer than patients with classic CJD.

- **variety**

see **sub-species.**

- **vascular plant**

plants that have a vascular system of vessels for conducting liquids. Include all flowers, trees, ferns etc.

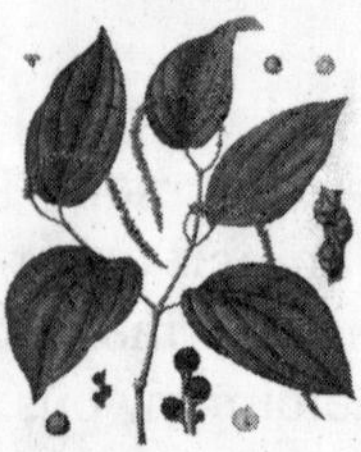

■ **vasectomy**
surgical cutting and sealing of part of male vas deferens to sterilise.

■ **vat pasteurization**
the heating of milk, while stirring in a vessel, to a temperature of at least 62.8°C (145°C) for 30 minutes.

■ **veal**
1. meat from the carcass of a young calf.
2. the meat from a calf or young beef animal. Male dairy calves are used in the veal industry. Dairy cows must give birth to continue producing milk, but male dairy calves are of little or no value to the dairy farmer. A small percentage are raised to maturity and used for breeding.
3. a calf (usually male) that is raised on milk and is intended to be used for meat at a young age. Veal meat is served at many restaurants and is very popular in cultural cuisine.
4. a calf fed for early slaughter.

■ **veal calf**
a calf is a young bovine of either sex that has not reached puberty (up to about nine months of age), and has a maximum live weight of 750 pounds.

■ **veal, 'bob'**
about fifteen percent of veal calves are marketed up to three weeks of age or at a weight of 150 pounds. These are called Bob Calves.

■ **veal, 'special-fed'**
the majority of veal calves are 'special-fed'. A veal calf is raised until about 16 to 18 weeks of age, weighing up to 450 pounds. They are raised in specially designed facilities where they can be cared for and monitored. Special, milk fed, and formula fed are the names

given to nutritionally balanced milk or soy based diets fed to calves. These diets contain iron and 40 other essential nutrients, including amino acids, carbohydrates, fats, minerals and vitamins.

- **vealer**
  calves fed (usually only milk) for early slaughter, usually less than 4 months and 350 pounds.

- **vector**
  the agent (e.g., plasmid or virus) used to carry new DNA into a cell.

- **vegan**
  a vegetarian whose diet excludes the use of all foods of animal origin.

- **vegetarian**
  person whose diet is restricted to foods of vegetable origin, including fruits, nuts and grains. Many vegetarians eat eggs and milk, but avoid all animal flesh.

- **verification**
  the use of methods, procedures, or tests by supervisors, designated personnel, or regulators to determine if the food safety system based on the HACCP principles is working to control identified hazards or if modifications need to be made.

- **vertebrates**
  animals with backbones (Fish, amphibians, reptiles, birds, mammals).

- **vertical integration**
  the integrating of successive stages of the production and marketing functions under the ownership or control of a single management organisation. For example, much of the broiler industry is highly vertically integrated in that processing companies own or control the activities from production and hatching of eggs, through the growth and feeding of the chickens, to slaughter, processing and wholesale marketing.

- **veterinary services**
  see **animal health care services.**

- **vfa (volatile fatty acids)**
  commonly used in reference to acetic, propionic, and butyric acids produced in the rumen of cattle, goats, and sheep, in the cecum of sheep,

the cecum and colon of swine, the colon of the horse, and the cecum of the rabbit.

- **video presentation**
training method with the use of a video.

- **video-assisted surgery**
surgery that is aided by the use of a video camera which projects and enlarges the tumour on a television screen. Also called video-assisted resection.

- **vinblastine**
belongs to the group of anticancer drugs known as vinca alkaloids.

- **vincristine**
a chemotherapy drug used in the treatment of cancer.

- **vindesine**
a drug used to stop the growth of tumour cells.

- **vinorelbine**
belongs to the group of anticancer drugs known as vinca alkaloids. Also known as Navelbine.

- **virion**
an elementary viral particle consisting of genetic material and a protein covering.

- **virology**
study of viruses.

- **virulence**
1. ability to infect or cause disease.
2. the degree of pathogenicity (ability of produce disease) of a microorganism as indicated by case fatality rates and/or its ability to invade the tissues of a host.

- **virus**
1. a submicroscopic organism that contains genetic information but cannot reproduce itself. To replicate, it must invade another cell and use parts of that cell's reproductive machinery.
2. microscopic organisms that cause infectious disease. In cancer therapy, some viruses may be made into vaccines that help the body build an immune response to and kill tumour cells.

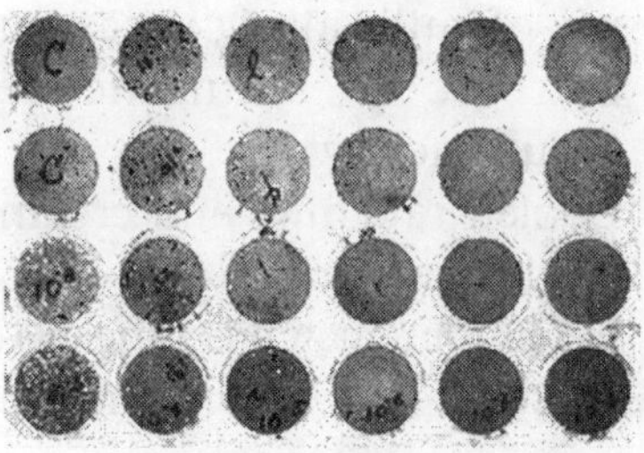

■ **viscosity**

viscosity of milk and milk products is important in determining the following

1. the rate of creaming
2. rates of mass and heat transfer
3. the flow conditions in dairy processes

Milk and skim milk, excepting cooled raw milk, exhibit Newtonian behaviour, in which the viscosity is independent of the rate of shear. The viscosity of these products depends on the following:

1. Temperature

cooler temperatures increase viscosity due to the increased voluminosity of casein micelles.

temperatures above 65° C increase viscosity due to the denaturation of whey proteins.

2. pH

an increase or decrease in pH of milk also causes an increase in casein micelle voluminosity.

Cooled raw milk and cream exhibit non-Newtonian behaviour in which the viscosity is dependant on the shear rate. Agitation may cause partial coalescence of the fat globules (partial churning) which increases viscocity. Fat globules that have under gone cold agglutination, may be dispersed due to agitation, causing a decrease in viscosity.

■ **viscus (plural, viscera)**

an internal organ of a human or animal.

■ **visera**

the internal organs of the body.

■ **visible identification**

a readily visible, numbering system attached to the animal that is used to identify the animal easily on test day.

■ **vitamin A**

nutrient added to low-fat and skim milk.

■ **vitamin D**

1. vitamin D is essential for absorption of calcium and phosphate particularly during childhood. It prevents rickets and helps maintain healthy teeth and bones. In excess, it can cause kidney stones. This condition is called Hypervitaminosis D.

2. nutrient that helps the body absorb calcium. It is often added to milk.

■ **vitamins**

1. micronutrients which are essential for good health and required by the body in small amounts. Generally, vitamins cannot be manufactured by the body and must be obtained from food.
2. vitamins are organic substances essential for many life processes. Milk includes fat soluble vitamins A , D, E, and K. Vitamin A is derived from retinol and ß -carotene. Because milk is an important source of dietary vitamin A, fat reduced products which have lost vitamin A with the fat, are required to supplement the product with vitamin A.

Milk is also an important source of dietary water soluble vitamins:
B1 - thiamine
B2 - riboflavin
B6 - pyridoxine
B12 - cyanocobalamin
niacin
pantothenic acid
There is also a small amount of vitamin C (ascorbic acid) present in raw milk but is very heat-labile and easily destroyed by pasteurisation.

■ **volatilisation**

the loss of gaseous materials, such as ammonia nitrogen, from animal manures to the atmosphere.

■ **vulnerable**

taxa believed likely to move into the endangered category in the naer future if the causal factors continue operating.

■ **vx-710**

an investigational drug that is used to help prevent resistance to cancer drugs.

■ **wader**

a general term for members of nine families of birds that overwinter in the intertidal zone in estuaries. Generally these are long.

■ **walk-through parlour**

upon completion of milking, cows walk through the front of the stall to exit.

■ **warfarin**

an anticoagulant drug. Also called a blood thinner.

■ **wash pen**
a corral or pen with a solid floor and permenent sprinklers or other washing devices for cleaning cows collectively prior to milking. Usually associated with a holding area.

■ **watchful waiting**
planned treatment in which the doctor follows your condition closely for any changes, without actual treatment, until symptoms appear. If the situation changes, curative treatments such as surgery or radiation can then be used.

■ **water chemistry**
chemical composition and characteristics of water (e.g. pH, amounof dissolved NPK, etc.).

■ **water quality objective**
a statutory target for water quality used to provide a common framework for dischargers and regulators.

■ **water table**
the upper level of the zone of groundwater saturation in permeable rocks.

■ **waterbody**
any land feature that holds water (e.g. lake, pond, river, etc.).

■ **watercourse**
a linear waterbody.

■ **waterlogging**
the character of a soil where drainage is impeded and the soil contains all the water it can absorb and more.

■ **waterway**
a navigable watercourse (e.g. river or canal).

■ **weighing**
procedure used by supervisors and dairy producers to determine the amount of milk given by a cow on test day.

■ **westphalian ham**
a German-style dry cured ham that is similar to Prosciutto, smoked, sometimes with juniper berries. Also called Westfalischer Schinken.

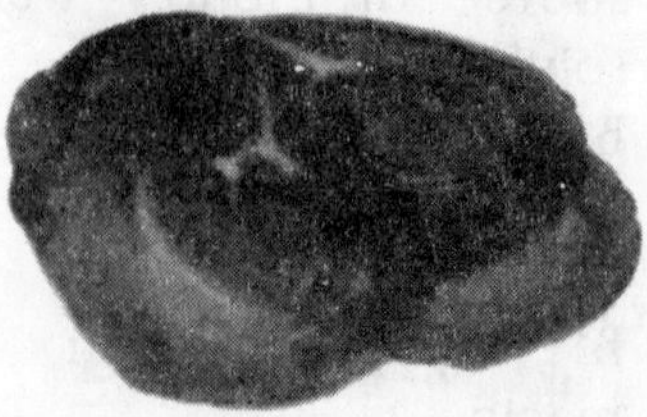

■ **wet solids**

condensed milk, skim milk, or whey may be referred to as wet solids, to distinguish from dry solids in the form of nonfat dry milk or dried whey powder.

■ **wetland**

general term for habitats and ecosystems that are dependent on large quantities of water to function (e.g. reedbeds, mosslands).

■ **whey**

1. watery part remaining after milk has curdled.
2. the residue from milk after removal of the casein and most of the fat, as occurs in cheese making. Although whey is 92% water it also contains 1% protein, together with all the milk lactose, water soluble vitamins and minerals.
3. the water and solids of milk that remain after cheese making and after the curd is removed. It contains about 93.3 percent water and 6.5 percent lactose, protein, minerals, enzymes; water–soluble vitamins and 0.2% fat.
4. the water and solids of milk that remain after the curd is removed. It contains about 93.5 percent water and 6.5 percent lactose, protein, minerals, enzymes, water-soluble vitamins and traces of fat.

■ **whey protein**

1. dairy source of protein (other than cassein), known for high levels of BCAA's and high nitrogen retention. Made from milk curd, whey protein is the Rolls Royce of proteins because it has a superior amino acid composition (including high levels of leucine, arguably the most important branched chain amino acid), superior biological value (meaning that more of what you eat gets digested and into your system), is very low in lactose (a milk sugar that most adults have difficulty digesting).
2. milk serum proteins. The major whey proteins in cow's milk are: alpha-lactalbumin and beta-lactoglobulin.

3. the proteins appearing in the supernatant of milk after precipitation at pH 4.6 are collectively called whey proteins. These globular proteins are more water soluble than caseins and are subject to heat denaturation. Native whey proteins have good gelling and whipping properties. Denaturation increases their water holding capacity. The principle fractions are ß-lactoglobulin, alpha-lactalbumin, bovine serum albumin (BSA) and immunoglobulins (Ig).

- **whipped butter**

sweet cream butter that has had air incorporated into it in order to increase its volume and make it easy to spread. Whipped butter cannot be substituted for butter in most recipes because its volume has been increased by air.

- **whipping cream**

a product sold as light whipping cream (30% to 36% milkfat) or heavy whipping cream (at least 36% milkfat). Whipping cream is not homogenised as this process interferes with the 'whipability' of the cream.

- **white blood cell**

white blood cells have a number of roles in the immune system, including antibody production, attacking and destroying cancer cells, and producing substances that kill cancer cells. Also called lymphocytes.

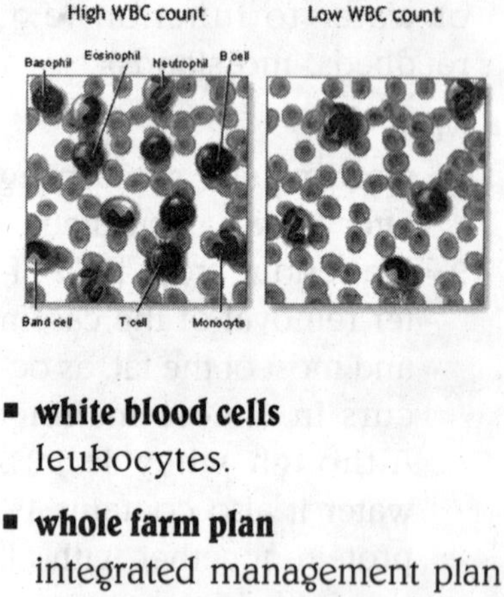

- **white blood cells**

leukocytes.

- **whole farm plan**

integrated management plan for a whole farm which incorporates physical characteristics, information on agricultural activities and environmental features. May include management objectives.

- **whole milk**

1. full cream milk = skim milk + cream.
2. milk which contains at least 3.25% milkfat and

8.25% nonfat solids. Almost all whole milk is fortified with vitamin D.

- **wholesale**
  the middle link in the food distribution chain. Wholesalers assemble relatively large quantities of product and resell in smaller lots to various users such as the food service trade, small retail food stores, jobbers, etc. Major, functions may include assembling, grading, warehousing and order taking and delivery customer service such as merchandising aids, credit, etc., also may be provided.

- **wild type**
  the form of an organism that occurs most frequently in nature.

- **wildfowl**
  ducks, swans and geese.

- **wildlife corridor**
  a linear feature (e.g. canal, hedgerow) that allows the movement of animals and plants between isolated sites.

- **wildlife site**
  a non-statutory designation of sites at the county/district level.

- **wildwood**
  the original woodland cover, relatively unaffected by human activity.

- **willow bark**
  white willow bark is a source of salicin, a chemical relative of asparin. The effects of willow bark are milk pain inhibition and increased blood flow to the skin and greater heat loss.

- **wing shoulder**
  a condition in which the shoulder joint is away from the rib structure and skeleton (much of the skin is tucked in behind it).

- **wingspan**
  the distance between the wing tips of any winged animal.

- **with calf**
  terminology designating a cow that is pregnant.

- **withdrawal time**
a 'withdrawal' period is required from the time antibiotics are administered until it is legal to slaughter the animal. This is so residues can exit the animal's system.

- **withdrawal time**
time required after the last drug treatment to lower drug residues to acceptable levels. These times are established using healthy animals according to label directions. An amount of time required following use of a medication in an animal before milk or meat can be entered into the human food supply.

- **worker**
among social insects such as ants, the workers are members of the colony that build the nest, fetch and store food and feed other members of the colony. They are usually sterile females.

- **workup**
the organised sequence of medical tests used to establish a diagnosis. A workup for diagnosis of kidney stones would include a urine examination and X-rays. A metabolic workup for stone prevention would refer to the blood and urine chemistry tests used to analyse chemical risk factors ultimately resulting in a diagnosis.

- **World Trade Organisation (WTO)**
the international organisation established by the Uruguay Round of multilateral trade negotiations to oversee implementation of the General Agreement on Tariffs and Trade and the agreements arising from the Uruguay Round, including the Uruguay Round Agreement on Agriculture.

- **wry tail**
tailhead set either to the right or left of centre.

- **xanthan gum**
a natural gum of high molecular weight produced by culture fermentation of glucose and used as a stabiliser in commercial food preparation.

- **xenobiotics**
synthetic chemicals believed to be resistant to environmental degradation. A branch of biotechnology

called bioremediation is seeking to develop biological methods to degrade such compounds.

- **xenotransplantation**
  the transplantation of living organs, cells or tissues from animals into humans.

- **x-ray**
  high-energy radiation used in low doses to diagnose diseases and in high doses to treat cancer.

- **x-ray crystallography**
  an essential technique for determining the three-dimensional structure of biological molecules. This information aids in the discovery of products that will interact with the biological molecule.

- **yearling**
  a male or female farm animal (especially cattle and horses) during the first year of life. Yogurt Fermented milk, lowfat milk or skim milk, sometimes protein-fortified. Lactobacillus bulgaricus and Streptococcus thermophilous are the fermenting bacteria. Fruit, flavours, and sugar may be added.

- **yearling mutton**
  meat from sheep that are generally slaughtered when they are older than 14 months of age, but were never a part of the breeding herd and show carcass maturity characteristics more advanced than those of lamb.

- **yearlings**
  animals that are between 12 and 24 months of age.

- **yeast**
  an organism that can grow and develop in the udder, causing mastitis.

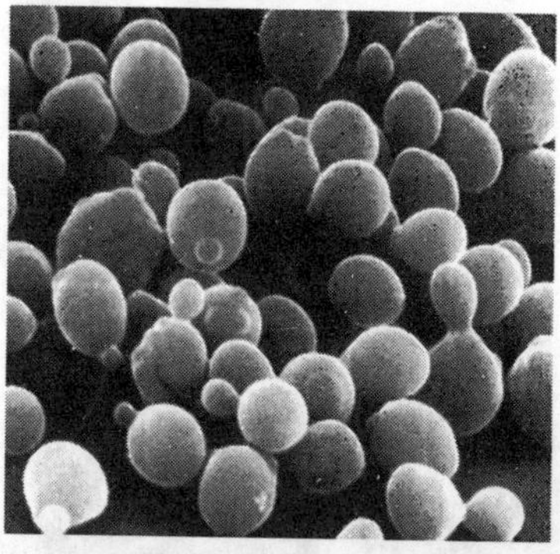

- **yeast-free diet**
  a diet most commonly associated with Candida or Candidiasis, in which certain foods are eliminated to reduce the growth of yeast within the body. Yeast containing products, such as

breads and other baked goods, fermented products, such as beer and alcohol, molds, and refined sugar are usually not acceptable for this kind of diet.

- **yield**
  the amount of product obtained from a fixed quantity of input material. Specific to pricing formulas used in the dairy industry, yield refers to how much butter, powder or cheese can be obtained from 100 pounds of milk. In California, the yields are 4.2 pounds of butter, 8.613 pounds of powder from milk testing 3.5% fat and 8.7% SNF. For milk testing 3.65% fat and 8.78% SNF, the cheese yield is 10.0.

- **yoghurt**
  fermented milk product.

- **yogurt**
  fermented milk, lowfat milk, or skim milk, sometimes protein– fortified. Fruit, flavours and sugars may be added. Milk solids content is commonly 15 percent. Most yogurt is high in protein and low in calories. Sometimes referred to as refrigerated yogurt to distinguish from frozen yogurt an ice cream–like product.

- **yohimbe**
  from the bark of an African tree, Yohimbe is a popular herb percieved as a stimulant and aphrodisiac. Yohimbe contains yohimbine, an alkaloid similar to caffeine in it's energising effects.

- **ziconotide**
  an investigational drug used in the treatment of chronic pain. Also known as SNX- I 1 1.

- **zinc oxide**
  a compound that may enhance immune ftmction, especially when administered by inhalation.

- **ZMA - zinc magnesium aspartate**
  this compound has been found to increase muscle strength.

- **zoonotic diseases**
  diseases that under natural conditions are communicable from animals to humans. Anthrax, Brucellosis, Psittacosis, Rabies, Tuberculosis, and Tularemia are example of zoonotic diseases. Brucellosis in livestock becomes undulant fever in humans.

Notes

Notes

Notes

Daya